patrick h. garrett

ANALOG I/O DESIGN

acquisition: conversion: recovery

reston publishing company, inc.
a prentice-hall company
reston, virginia

Library of Congress Cataloging in Publication Data

Garrett, Patrick H
  Analog I/O design.

  Includes bibliographies and index.
  1. Analog-to-digital converters. 2. Digital-to-analog
converters. 3. Computer interfaces. I. Title.
TK7887.6.G37   621.3819'596   80−28435
ISBN 0−8359−0208−0

Editorial/production supervision and interior design
by Norma M. Karlin
Manufacturing buyer: Ron Chapman

10 9 8 7 6 5 4 3 2 1

PRINTED IN THE UNITED STATES OF AMERICA

contents

chapter **7**   *DEVICES FOR DATA CONVERSION*   *161*

chapter **8**   *DATA CONVERSION SYSTEMS*   *191*

chapter **9**   *SIGNAL RECOVERY AND DISTRIBUTION*   225

PREFACE

Analog instrumentation is essentially associated with measurement, control, test, and computer interfacing systems. The recent growth in microcomputer applications, however, is imposing increased cost and performance accountability on the design of analog systems. This task is complicated by the fact that analog systems involve a diversity of subjects whose organization into specific designs usually requires a custom approach.

This book provides a comprehensive approach to the design of analog systems primarily for computer input/output applications. The nine chapters are organized to provide a gradual and logical exposition. Novel topics developed include a linear filtering equivalent to matched filtering for signal-quality upgrading, and a quantitative signal-recovery method which when coordinated with an input analog-to-digital conversion system permits selection of the minimum sample rate for the output-signal resolution of interest. An overall theme is the optimizing of designs, as by limiting signal spectral occupancy of filter passbands to minimize filter amplitude error, and by minimizing noise in low-level signal-measurement circuits. Presented throughout are component and system error analyses to permit evaluation of overall input-to-output system accuracy and thereby the realization of a unified design approach for analog systems.

The text is intended for a one-term course on analog instrumentation and computer interfacing at the senior and first-year graduate level and as a reference for practitioners on the detailed design of analog I/O systems. The many examples and results included provide a practicality that a more formal approach might lack, and the book emphasizes clarity and usefulness for design applications rather than rigorous proofs. It is intended both to demonstrate the continuing requirement for analog instrumentation and to communicate some of the possibilities for effective analog system design.

Pat Garrett

introduction
to analog
systems

1-0 INTRODUCTION

Analog systems are used primarily in measurement, data-conversion, control-system, and signal-processing applications. A digital processor also is frequently involved, requiring instrumentation for the acquisition of sensor data and for output signal recovery and distribution operations. This is common in industrial process automation, data logging, laboratory and biomedical applications, and aerospace systems. Linear devices such as amplifiers and conversion products represent the basic elements of these systems for activation and signal-flow purposes. Sales of these components are experiencing a sustained growth, indicating their continued vitality.

Chapter 1 offers an orientation to the nature and scope of analog systems. This preview serves as an introduction to the diverse subjects that must be treated, including signal analysis, linear electronics, sampled-data considerations, and practical implementation requirements. Increasingly, performance and economic considerations require that analog systems realize the necessary accuracy without overdesign. Accordingly, this book presents a systems engineering design approach, involving comprehensive error analysis to accommodate signal parameters, component accuracies, and system factors. This approach yields a useful quantitative measure of overall performance and indicates where improvement is productive in achieving the desired accuracy.

1-1 SENSORS AND SIGNALS

The science of measurement has supported technology since its beginning, and a technology-intensive society has in turn imposed a large superstructure on the science of measurement. Measurement is an inexact science requiring the use of reference standards, which are involved either directly or indirectly in all measurements, more directly as the requirement for accuracy increases. For example, highly accurate measurements are generally obtained by the comparison method, such as the voltage measurement potentiometer employing a standard cell that is traceable to the National Bureau of Standards. More typical, however, is the periodic comparison and correction of a sensor measurement against a transfer standard that embodies some calibration procedure. Even gross measurements, such as the use of sensors to detect high or low limits in a process, require an initial reference comparison and calibration.

A *transducer* is a device capable of transferring energy between two systems, as in the conversion of thermal into electrical energy by the Seebeck-effect thermocouple. Table 1-1 lists various electrical transducers that illustrate the wide range of operating principles involved and the signal-conditioning operations necessary to convert the transducer outputs into useful electrical quantities. Basic transducer elements are presented in Figure 1-1; the details of these operations are developed in Chapters 5 and 6. Since the transducer output-signal amplitude normally corresponds to some measurand of interest, such as temperature or pressure, we want to acquire a true amplitude

TABLE 1-1
ELECTRICAL TRANSDUCER AND INTERFACING OUTPUTS

MEASURAND	SENSOR	PRINCIPLE	CIRCUITS
Temperature	Dissimilar metals	Thermoelectric	Reference junction
Pressure	Diaphragm	Potentiometric	DC excitation
Light intensity	CdS cell	Photovoltaic	Logarithmic compression
Strain	Piezoresistors	Bridge circuit	Linearization
Level	Sonic	Ultrasound	Timing measurements
Flow	Turbine	Tachometer	Voltage scaling
Acceleration	Ceramic crystal	Piezoelectric	Charge amplification
Velocity	Hot wire	Anemometer	Current sensing
Displacement	LVDT	Reluctive	AC excitation
pH	Ion-selective electrode	Nernst relationship	Electrometer

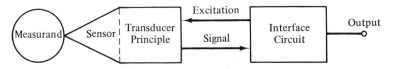

Figure 1-1 Transducer Circuit Elements

measurement. This can be achieved through an optimum system design that minimizes both internal and external interference sources and accounts for essential component and system errors. Measurement uncertainty due to noise is a common source of error, which can frequently be reduced by filtering, but this requires information about signal spectral occupancy requirements. Consequently, determination of the signal bandwidth is an initial requirement in the design of instrumentation systems. Other transducer specifications of interest include amplitude response range, linearity error, and calibration requirements.

Most analog signals vary in magnitude over time, and this variation conveys the signal's information content. In order to amplify or otherwise process this signal, however, the minimum bandwidth that accurately represents the signal spectral occupancy must be provided. This spectral occupancy, essentially a function of the rate of change of the signal, can be satisfactorily determined by the intelligent use of an oscilloscope with a priori knowledge of the frequency spectrum of typically encountered signals. This is a practical approach because of the usual availability of an oscilloscope and the small number of classifications into which instrumentation signals generally may be grouped.

The Fourier spectrum is an efficient method of representing the information bandwidth requirements of a signal. We arrive at this spectrum by acknowledging that any arbitrary time-domain periodic signal may be expressed by sinusoidal components whose amplitudes and phases, when summed, equal the instantaneous signal amplitude and polarity. Equation (1-1) is a discrete Fourier expression for a periodic signal whose period T equals $1/f_1$, where a_0 represents the dc component, a_n the even, and b_n the odd harmonic components. Of particular interest is the rate at which these harmonic sinusoidal components diminish with increasing frequency, thereby determining the required signal spectral occupancy.

$$f_{(t)} = a_o + a_1\cos(2\pi f_1 t) + a_2\cos 2(2\pi f_1 t) + \cdots$$
$$+ b_1\sin(2\pi f_1 t) + b_2\sin 2(2\pi f_1 t) + \cdots \qquad (1\text{-}1)$$

We can gain insight into this matter by examining the waveforms of Figure 1-2 and their corresponding frequency spectrums. Note that the sharp transitions of the trapezoidal and sawtooth waves result in greater harmonic amplitudes than the triangular wave. Providing a flat-amplitude signal passband out to approximately the 10% harmonic amplitude frequency will provide for an adequate signal representation, accommodating the typical -20-dB/decade rolloff of most signals with complex harmonic content. Application of Table 1-2 to oscilloscope signal observations therefore permits

Figure 1-2 Instrumentation Signal Waveform Classifications

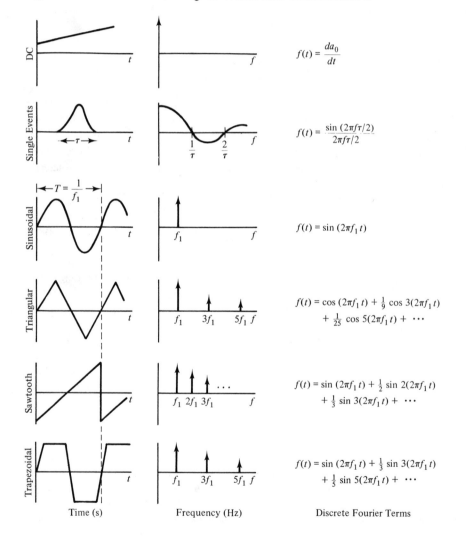

DC

$$f(t) = \frac{da_0}{dt}$$

Single Events

$$f(t) = \frac{\sin (2\pi f \tau/2)}{2\pi f \tau/2}$$

Sinusoidal

$$f(t) = \sin (2\pi f_1 t)$$

Triangular

$$f(t) = \cos (2\pi f_1 t) + \tfrac{1}{9} \cos 3(2\pi f_1 t) + \tfrac{1}{25} \cos 5(2\pi f_1 t) + \cdots$$

Sawtooth

$$f(t) = \sin (2\pi f_1 t) + \tfrac{1}{2} \sin 2(2\pi f_1 t) + \tfrac{1}{3} \sin 3(2\pi f_1 t) + \cdots$$

Trapezoidal

$$f(t) = \sin (2\pi f_1 t) + \tfrac{1}{3} \sin 3(2\pi f_1 t) + \tfrac{1}{5} \sin 5(2\pi f_1 t) + \cdots$$

Time (s)   Frequency (Hz)   Discrete Fourier Terms

TABLE 1-2
SIGNAL CLASSIFICATION AND BANDWIDTH REQUIREMENT

SIGNAL TYPE	MINIMUM BANDWIDTH
DC	Time rate change
Sinusoidal	1/period
Single events	2/width
Periodic triangular	3/period
Periodic sawtooth	10/period
Periodic trapezoidal	10/period

classification and signal-bandwidth determination. For signals that only approximate the waveforms shown, the more severe classification should be assumed. That is, the waveform having the sharpest transitions that is a reasonable facsimile of the signal should be chosen to insure adequate bandwidth. Our interest here is in using the results of Fourier analysis for typically encountered instrumentations signals, rather than pursuing the details of this mathematical signal-analysis method.

1-2 INSTRUMENTATION AND CONVERSION CIRCUITS

It can now be appreciated that measurement signals are an important element in analog instrumentation, and that the achievement of an accurate measurement requires attention to a number of complicated considerations. This section presents a broad outline of electronic analog instrumentation functions and circuits that, in general, provide the translation and manipulation required by the various signals. Functions are grouped into four classifications in Table 1-3; an important example of each is presented in Figures 1-3 through 1-6, and a chapter of the book is devoted to the development of each classification.

Operational amplifiers are generally the first active devices encountered at the input of analog systems, where they are frequently called *preamplifiers*. Progress in analog instrumentation and signal processing has been especially

TABLE 1-3
ANALOG INSTRUMENTATION FUNCTIONS

ACQUISITION	CONVERSION	RECOVERY	PROCESSING
Sensor	Multiplexer	Digital-analog	Averaging
Amplification	Sample-hold	Reconstruction	Computation
Filter	Analog-digital	Distribution	Conditioning
Isolation	Programmable	Transmission	Logarithmic

rapid since the introduction of the Fairchild 709 monolithic integrated-circuit operational amplifier in 1965. Operational amplifiers do offer less precision and long-term stability than instrumentation amplifiers, however. Differential dc amplifiers were available for instrumentation applications before this time, of course, and extend back to the early vacuum-tube Philbrick amplifiers commercially available late in the 1940s. Before the introduction in the early 1960s of the very-high-gain devices permitting a stabilizing large negative feedback, the output voltage drift of these dc amplifiers limited their utility and widespread application.

A primary utility of the differential-input operational amplifier is its ability to amplify dc signals stably and ac signals simultaneously without phase shift because of direct-interstage coupling. The availability of an inverting and noninverting input is of particular interest for instrumentation applications because of the common-mode interference rejection this amplifier input arrangement offers. For interference coupled to both inputs, equal but opposite polarity amplification is provided, achieving a high degree of interference rejection. However, as discussed in Chapter 3, the primary purpose of the differential-input stage that provides this rejection is to ensure good dc stability.

Active filter networks using operational amplifiers permit the realization of stable, yet inexpensive frequency-selective networks from dc to about 100 kHz. Filtering at the lower instrumentation frequencies has always been a problem with passive filters because of the required L and C values and large inductor losses. Although the history of electric wave filters extends over half a century, the identification of stable and efficient active networks with emphasis on their application has occurred only in the past decade. This development has been especially attractive for instrumentation applications because it has made possible high-performance transducer-amplifier-filter signal-acquisition circuits for upgrading the quality of measurement signals. Figure 1-3 shows

Figure 1-3 Signal Acquisition and Conditioning

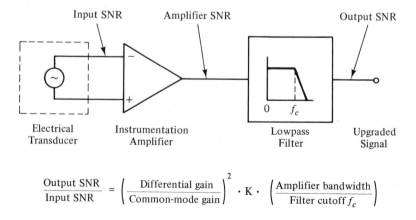

$$\frac{\text{Output SNR}}{\text{Input SNR}} = \left(\frac{\text{Differential gain}}{\text{Common-mode gain}}\right)^2 \cdot K \cdot \left(\frac{\text{Amplifier bandwidth}}{\text{Filter cutoff } f_c}\right)$$

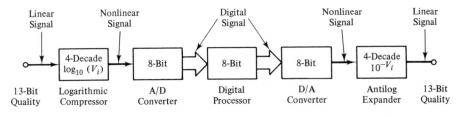

Figure 1-4 Logarithmic Signal Processing

an acquisition and conditioning circuit with the canonical equation describing the signal-quality improvement in terms of signal-to-noise ratio (SNR). The simplified processing gain relationship of equation (1-2) is based on the derivation (in Chapter 6) of linear filtering efficiency K in relationship to optimum matched filtering for instrumentation signals. This is especially important to low-level signal conditioning where the improvement in SNR provided by equation (1-2) is frequently 10^{10}. Signal-conditioning design examples also are presented in Chapter 6, supported by the basic building-block functions developed in the earlier chapters of the book. Notice that signal conditioning is essentially concerned with operations on measurement signals, whereas signal processing addresses a broader classification of signal operations.

Signal-processing operations generally involve signal manipulations following improvement of quality. These operations include averaging and linearization of transducer signals, computational operations for data reduction, and logarithmic signal compression and expansion for binary wordlength reduction in data-conversion systems. The latter operation, illustrated in Figure 1-4, is especially useful for processing high-resolution transducer signals by truncated binary word-length digital processors. If the data eventually are to be reconstructed in analog form, an antilog operation following output D/A conversion is required to restore signal linearity. This method results in a constant fractional error throughout the signal dynamic range at the expense of high resolution at any point within that range owing to the exponential distortion of the signal over its range. SNR can also be improved with logarithmic compression, which may be of interest in remote data-acquisition applications, and A/D conversion speed is increased because of the shorter binary wordlength.

Analog computation of process data is useful when the reduction of sensor data from several inputs is required, such as in the dedicated calculation of mass flow rate. The rationale for an analog realization is that the transducer inputs are available in analog form, and perhaps ten mathematical operations can be provided at an overall accuracy of 0.5% to 1% (about seven-bit quality) for a cost below that of a microcomputer-based system that also includes an analog input subsystem. This is made possible by the computation

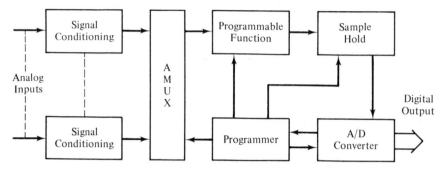

Figure 1-5 Multiplexed Data Conversion

efficiency of devices such as the multifunction module and analog multiplier/dividers. Such mechanizations can also serve as a local analog data concentrator, whose output may then be transmitted to a remote computer.

Sensor-based data-acquisition systems for industrial and laboratory applications contain assemblies of components for signal-acquisition, conditioning, and conversion purposes. It is common practice for reasons of economy to share components in multiplexed data-conversion systems and to choose from among various available approaches and conversion methods in the implementation of a system according to the specific task. For example, users can buy components and construct a system, buy modular subsystem assemblies and provide the requisite interfacing and support, or purchase a complete standalone system. Chapter 8 explores these approaches, using detailed design examples for both low- and high-data-rate systems.

Data-conversion components are presented in detail in Chapter 7, ranging from analog multiplexers to the various available analog-to-digital conversion methods. These range from the very-high-speed eight-bit parallel

Figure 1-6 Output Signal Recovery

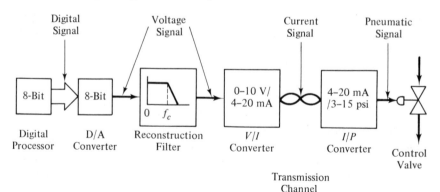

conversion 1007 device from TRW capable of 35 million conversions per second and intended for video data applications, to an austere voltage-to-frequency A/D converter capable of 39 eight-bit conversions per second intended for low-data-rate applications. Data-conversion devices are primarily constructed of analog components, and their design into systems is essentially an analog task. Figure 1-5 illustrates a conventional data-conversion system.

A fundamental consideration in the design of data-conversion systems is specification of the sample rate, which depends upon several factors, including signal aliasing, but is dominated by resolution requirements. The resolution provided by step-interpolator-represented sampled data depends upon the sample rate and signal rate of change. Sampled data are almost always recovered to proceed to their destination after digital processing by means of an output reconstruction process. Chapter 9 discusses signal-recovery smoothing methods, noting that practical techniques typically offer two to three bits increase in output resolution above step-interpolator-represented data. Consequently, sample rate should be selected for a data conversion system to achieve the recovered output accuracy of interest. Figure 1-6 describes an output signal channel including signal reconstruction, transmission, and actuation.

1-3 SYSTEM DESIGN CONSIDERATIONS

The application of systems engineering methods to the design of instrumentation systems helps us to evaluate the numerous tradeoffs necessary and to arrive at a preferred set of options from among available alternatives. For example, three budgets — system error, timing, and cost — often must be optimized. Furthermore, a cost-benefit analysis is usually of interest in order to achieve the best mix of performance, cost, and system reliability. These criteria are described in Figures 1-7(a) and (b), where weighting functions are assigned according to merit appropriate to the mission of the system.

Performance evaluation is usually based on figures of merit, selected for their appropriateness to the tasks the system is to perform. For example, evaluation of data-conversion systems may be based on cost and throughput. In a wider sense, a general figure of merit for a data-acquisition system is its ratio of information to data input; the need is for more information from less data. For example, Figure 1-8 presents an analog signal $x(t)$ that provides a detailed point-by-point data record in time. A judicious choice of signal-processing functions, however, can summarize the data into a more efficient form of information. We might choose data averaging avg $|x(t)|$, generation of a histogram $g(x)$ of the signal-amplitude distribution, or the amplitude-squared spectrum $[x(t)]^2$ as a function of frequency. This latter power-density spectrum provides the instrumentation designer with insight into the process represented by the signal and its response.

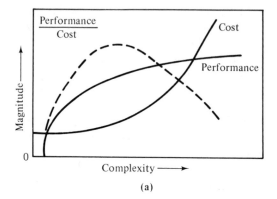

(a)

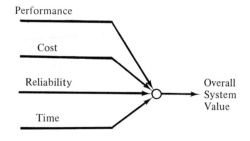

(b)

Figure 1-7 Cost-Benefit Criteria

Figure 1-8 Analog Signal Record

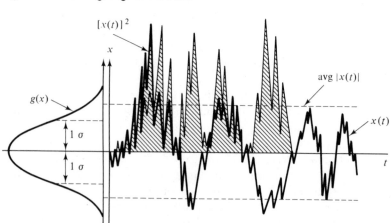

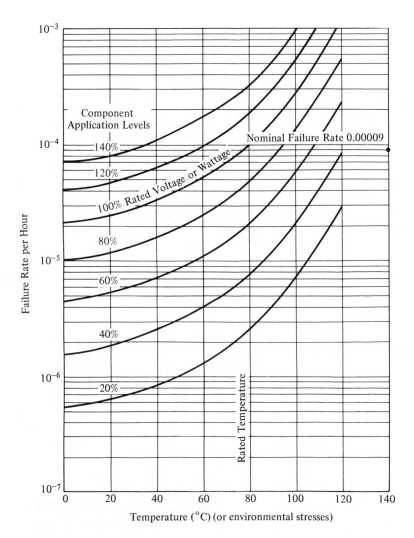

Figure 1-9 Failure-Rate Derating Curves for Electronic Components
(Igor Bazovsky, Reliability Theory and Practice, © 1961, p.149. Adapted by Permission of Prentice-Hall, Inc. Englewood Cliffs, N.J.)

   In the design of any system it is useful to bracket a specification with both overdesigned and austere solutions in order to realize an optimum mechanization. Also, analog systems are mostly series connected with little redundancy involved. Therefore, overall reliability makes component derating and design simplification important. Operation at stress levels below component ratings provides substantial reliability enhancement, as shown in Figure 1-9. Component failure rates are essentially a function of thermal and electrical stresses, which Arrhenius' law reflects by these curves. Derating components by about one-third in both temperature and operating voltage typically reduces failure rate by an order of magnitude. Reliability enhancement is also available from the specification of premium components, such as from the military reliability grade of unscreened mil temp. This category offers both linear and digital active devices with twice the reliability of commercial-grade components. Concern is primarily with active devices, since passive components operated within their ratings exhibit much lower failure rates.

A quantitative performance measure developed throughout the text is a comprehensive error analysis, which includes deterministic errors for the electronic devices and circuits, factors associated with their design into systems, and the effects of internal and external influences. A description of these error sources for analog I/O systems is presented in Table 1-4. Predominant are errors related to the effect of sample rate on resolution, component errors resulting from temperature variation, and the effect of noise on signal quality. Equations (1-3) and (1-4) show that if these error sources are uncorrelated, which is a reasonable assumption, then they may be combined in an rms

TABLE 1-4
ANALOG I/O SYSTEM ERROR SOURCES

SYSTEM	ERROR SOURCE	DESCRIPTION
Acquisition	Transducer	Nonlinearity as percent of full scale
	Interface	Termination circuit error
	Amplifier	Component error generally referred to its input
	Filter	Amplitude error due to gain and phase
	Signal quality	Signal accuracy following upgrading
Conversion	Multiplexer	Component and transfer error
	Sample-hold	Component and acquisition error
	Aperture	Uncertainty in acquired signal amplitude
	Resolution	Step-interpolator data amplitude error
	Aliasing	Bandwidth, residual-noise, and sample-rate dependent
	Sin x/x	Attenuation due to sample rate
	A/D	Component error including quantization
Recovery	D/A	Component error including differential nonlinearity
	Sin x/x	Attenuation due to update rate
	Filter	Amplitude error due to gain and phase
	Aliasing	Update-rate dependent
	Resolution	Reconstructed-output-signal amplitude error

fashion. This operation is valid because the cross-product terms are zero as shown, and it demonstrates that larger errors dominate the error budget ϵ_{a+b+c}, thereby indicating where improvement is productive in achieving the overall accuracy of interest. Consequently, system performance can be measured through determination of its input-to-output accuracy, which also provides a unified design approach for analog systems.

$$
\begin{aligned}
\epsilon_{a+b+c}^2 &= (\epsilon_a + \epsilon_b + \epsilon_c)^2 \qquad\qquad (1\text{-}3)\\
&= \epsilon_a^2 + 2\epsilon_a^{0}\epsilon_b + \epsilon_b^2 + 2\epsilon_a^{0}\epsilon_c + \epsilon_c^2 + 2\epsilon_b^{0}\epsilon_c\\
&= \epsilon_a^2 + \epsilon_b^2 + \epsilon_c^2
\end{aligned}
$$

$$
\epsilon_{a+b+c} = \sqrt{\epsilon_a^2 + \epsilon_b^2 + \epsilon_c^2} \qquad\qquad (1\text{-}4)
$$

REFERENCES

1. N. A. ANDERSON, *Instrumentation for Process Measurement and Control*, Chilton, Radnor, Pa., 1972.

2. I. BAZOVSKY, *Reliability Theory and Practice*, Prentice-Hall, Englewood Cliffs, N.J., 1965.

3. H. CHESTNUT, *Systems Engineering Methods*, John Wiley & Sons, New York, 1967.

4. P. H. GARRETT, *Analog Systems for Microprocessors and Minicomputers*, Reston, Reston, Va., 1978.

5. M. D. MESAROVIC, *Views on General Systems Theory*, John Wiley & Sons, New York, 1964.

6. *RADC Reliability Notebook*, Publication RADC-TR-65-330, Rome Air Development Center, Rome, N.Y., 1965.

linear
electronics

2-0 INTRODUCTION

This chapter is concerned with transistor devices, circuits, and techniques that comprise the electronic components of analog systems. Most linear electronic circuits are subjected to small-signal perturbations that only slightly vary the associated current and voltage levels. Digital circuits, on the other hand, are normally subjected to large-signal changes. Consequently, linear circuit design and analysis benefit from the use of small-signal models. The h-parameter model is standard for low-frequency small-signal circuits employing bipolar transistors, and the hybrid-π model is generally used for JFET and MOSFET circuits.

Our discussion employs a simplified form of these models to compare the voltage gain and terminating impedances of elemental transistor circuits using identical load impedances and voltages; we also consider the temperature limitations of semiconductor devices. These basic circuits are then included in circuit structures important to linear electronics, such as constant-current sources and active-load high-gain amplifier stages. This progression is extended to differential dc amplifiers, developing their key parameters and errors. All this provides the basis for understanding the capabilities and limitations of the operational amplifier, the universal active device of analog instrumentation systems. We also discuss high-frequency aspects of the operational amplifier, including phase compensation to prevent device oscillation, gain-bandwidth interactions in multistage circuits, distortion relationships, and output loading effects. This chapter also presents frequently encountered signal processing operations including signal averaging, computation for data-reduction purposes, and logarithmic companding both for signal transmission and for the accommodation of high-resolution signals by truncated-wordlength digital processors. The availability of capable linear devices for implementing these circuits continues to promote the utility of analog realizations for many functions.

2-1   PN-JUNCTION TEMPERATURE CHARACTERISTICS

The pn junction is the basic semiconductor device in electronic circuits; among its forms are diodes and bipolar and JFET transistors. One must appreciate here that the availability of the free carriers that result in current flow in a semiconductor is essentially a function of applied thermal energy. At room temperature, taken as $17°C$ ($290°K$ above absolute zero), there is abundant energy to liberate the valence electrons from a semiconductor lattice structure. These free carriers can then drift under the influence of an external applied potential. However, this current flow is a function of the thermal energy and not of the applied voltage; this thermal phenomenon accounts for the negative temperature coefficient exhibited by semiconductors (increasing current with increasing temperature).

The primary variation associated with reverse-biased pn junctions is the change in I_s, reverse saturation current. I_s increases 7 percent per degree centigrade rise for silicon, doubling every $10°C$, and has a nominal value in the 10^{-9}-A range at room temperature in equation (2-1). Forward-biased pn junctions exhibit a decreasing junction potential having a typical value of -2.5

Figure 2-1 PN-Junction Temperature Dependence

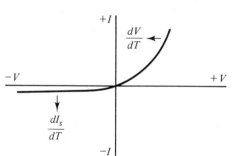

mV per degree centigrade rise described by equation (2-2), where T is in degrees centigrade. This dV/dT temperature dependence, determined by the difference between the expected silicon pn junction potential V of 0.6 volt and the temperature dependence of I_s, later is shown to be a primary contributor to the error budget in analog instrumentation systems. Figure 2-1 shows the pn junction volt-ampere relationship mathematically described by equation (2-3). The volt equivalent of temperature, $V_T = T°K/11,600$, is a variable in both of these equations, has a nominal value of 25 mV at room temperature, and represents the effect of the distribution of energy levels in the semiconductor.

$$\frac{dI_s}{dT} = -I_s \cdot \frac{d(\ln I_s)}{dT} \quad \text{A/°C} \tag{2-1}$$

$$\frac{dV}{dT} = \left(\frac{V}{T} - V_T \cdot \frac{d(\ln I_s)}{dT} \right) \quad \text{V/°C} \tag{2-2}$$

$$I = I_s \left[\exp \left(\frac{V}{V_T} \right) - 1 \right] \quad \text{A} \tag{2-3}$$

2-2 SMALL-SIGNAL LOW-FREQUENCY TRANSISTOR CIRCUITS

Earlier eras required the use of rather elaborate models to account for the behavior of transistors when their various anomalies were considered. However, with improvements in both discrete and monolithic devices many of these departures from the ideal have become negligible, permitting simplification of linear circuit design and analysis. For example, the h-parameter model that is standard for small-signal low-frequency bipolar transistors can be simplified to that shown in Figure 2-2. The collector output conductance h_{oe} is

Figure 2-2 Approximate Low-Frequency Transistor Models

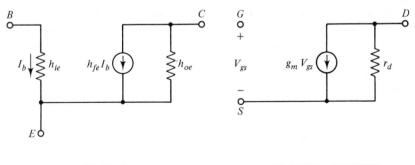

(a) Bipolar   (b) JFET and MOSFET

TABLE 2-1
LOW-FREQUENCY TRANSISTOR AMPLIFIER RELATIONSHIPS

	Common Emitter	Common Collector	Common Base	Common Source	Common Drain
Input Impedance, R_i	h_{ie}	$h_{ie} + h_{fe}R_L$	$\dfrac{h_{ie}}{h_{fe}}$	$> 10^9\,\Omega$	$> 10^9\,\Omega$
Voltage Gain, A_v	$\dfrac{-h_{fe}R_L}{h_{ie}}$	$\dfrac{h_{fe}R_L}{h_{ie} + h_{fe}R_L}$	$\dfrac{h_{fe}R_L}{h_{ie}}$	$\dfrac{-g_m r_d R_L}{r_d + R_L}$	$\dfrac{g_m R_L}{1 + g_m R_L}$
Output Impedance, R_o	R_L	$\dfrac{h_{ie}}{h_{fe}}$	R_L	$\dfrac{r_d R_L}{r_d + R_L}$	$\dfrac{1}{g_m}$

normally very small and usually neglected in circuit calculations. Similarly, the modified hybrid-π model useful for representing JFET and MOSFET transistors is simplified by the removal of various interelectrode capacitances. Bipolar, JFET, and MOSFET transistors are the primary electronic devices used in analog instrumentation circuits, and they provide a foundation that we can build upon by reviewing their basic gain, impedance, and circuit relationships.

Bipolar transistors represent the first three amplifier circuits tabulated in Table 2-1. It is common practice to present the three possible connections in terms of common-emitter h-parameters because this is the connection most frequently used. Figure 2-3 shows elemental circuits for the three connections of an npn transistor amplifier having a current gain h_{fe} of 100 and input resistance h_{ie} of 1 k, with the base-emitter junctions forward biased and collector-base junctions reverse biased. A 1-k load resistor R_L is used for all three circuits, the endpoints of the load lines being determined by the dc supply voltage and $(12\text{ V} - V_{ce})/R_L$. Fixed bias is used for simplicity to place the quiescent Q-points at the midpoint of the transfer characteristic for each circuit as shown. The choice of this bias resistor is based on the supply voltage and a transistor input base current of 0.06 mA. Note that the input impedance R_i is defined looking directly into the transistor base and is exclusive of the input circuit and bias network. The blocking capacitor C provides an ac signal path for V_s without upsetting the dc bias for the three circuits.

Comparable circuits are implemented with the high-input-impedance JFET transistor in Figure 2-4. These stages are self-biased by the 1-k source resistor at a quiescent gate-to-source voltage V_{gs} of -0.5 V. The selection of an appropriate JFET quiescent Q-point for a given transconductance g_m and drain resistance r_d is determined by considerations similar to those for bipolar

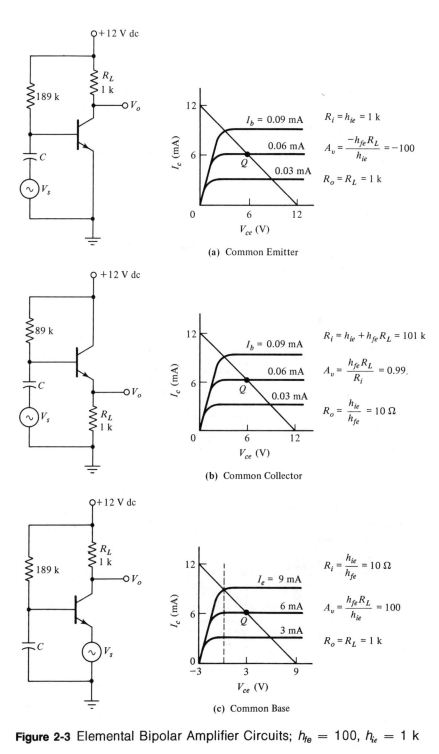

(a) Common Emitter

$R_i = h_{ie} = 1$ k

$A_v = \dfrac{-h_{fe} R_L}{h_{ie}} = -100$

$R_o = R_L = 1$ k

(b) Common Collector

$R_i = h_{ie} + h_{fe} R_L = 101$ k

$A_v = \dfrac{h_{fe} R_L}{R_i} = 0.99$

$R_o = \dfrac{h_{ie}}{h_{fe}} = 10\ \Omega$

(c) Common Base

$R_i = \dfrac{h_{ie}}{h_{fe}} = 10\ \Omega$

$A_v = \dfrac{h_{fe} R_L}{h_{ie}} = 100$

$R_o = R_L = 1$ k

Figure 2-3 Elemental Bipolar Amplifier Circuits; $h_{fe} = 100$, $h_{ie} = 1$ k

19

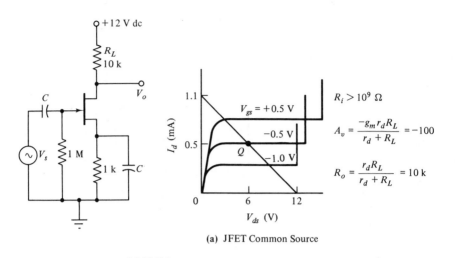

(a) JFET Common Source

$R_i > 10^9 \ \Omega$

$A_v = \dfrac{-g_m r_d R_L}{r_d + R_L} = -100$

$R_o = \dfrac{r_d R_L}{r_d + R_L} = 10 \text{ k}$

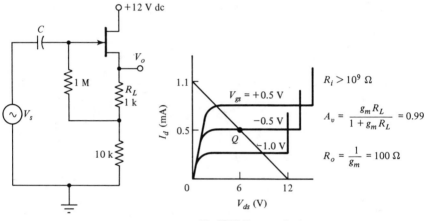

(b) JFET Common Drain

$R_i > 10^9 \ \Omega$

$A_v = \dfrac{g_m R_L}{1 + g_m R_L} = 0.99$

$R_o = \dfrac{1}{g_m} = 100 \ \Omega$

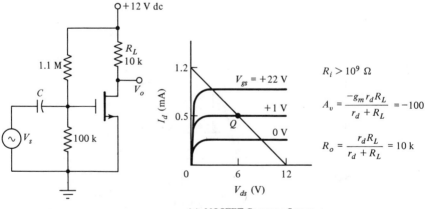

(c) MOSFET Common Source

$R_i > 10^9 \ \Omega$

$A_v = \dfrac{-g_m r_d R_L}{r_d + R_L} = -100$

$R_o = \dfrac{r_d R_L}{r_d + R_L} = 10 \text{ k}$

Figure 2-4 Elemental FET Amplifier Circuits; $g_m = 10^{-2}$ A/V, $r_d = 1$ M

devices. These include provision for the maximum undistorted output signal swing about the Q-point, and biasing against device variation by means of the -0.5 V of self-bias developed across the 1-k source resistor from the 0.5 mA of current flow through it. The 1-M resistor completes the circuit to the gate for this bias voltage, and the gate pn junction must always remain reverse biased. Bias stabilization is not included in the bipolar circuits of Figure 2-3; it is addressed in the subsequent development of a common-emitter amplifier with an active load. The enhancement-mode MOSFET is biased by a fixed-bias resistive voltage divider at $+1.0$ V to provide 0.5 mA of drain current. This device exhibits characteristics similar to those of the JFET transistor, the primary difference being that its gate is a capacitive element rather than a reverse-biased pn junction. The arrow directions for all of the FET transistors shown describe n-channel devices. For an extended review of transistor circuits and biasing refer to Reference 1. From a comparison of the bipolar and FET transistor circuits of Figures 2-3 and 2-4, it is evident that the FET configurations offer a much higher input impedance. This is primarily of value in preventing attenuation of the input signal due to the voltage-divider effect that a large series source resistance would impose, which would be internal to the signal source V_s. This is simply the loading effect on the signal source due to amplifier input impedance R_i. The common-emitter and common-source amplifiers offer high gain, but with low input impedance for the bipolar circuit. The common-collector and common-drain circuits are useful as input and output buffer stages, both offering high input-termination and low driving-point impedances. The common-base bipolar amplifier is of limited utility, however, because of its very low input impedance.

A bipolar transistor structure frequently used in linear electronic circuits is the constant-current source or sink shown in Figure 2-5. The source circuit

Figure 2-5 Constant-Current Source and Sink

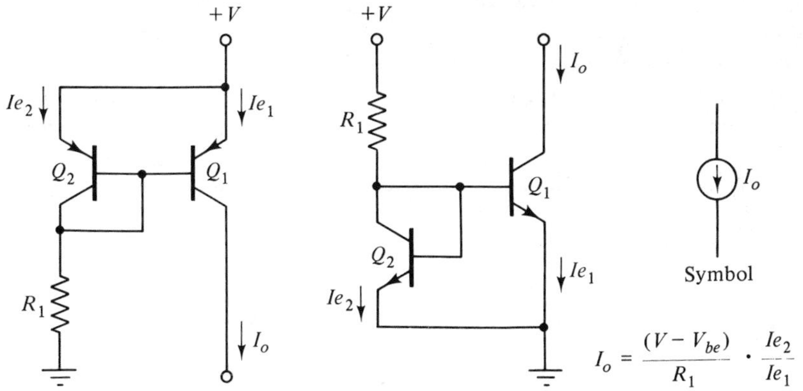

$$I_o = \frac{(V - V_{be})}{R_1} \cdot \frac{Ie_2}{Ie_1}$$

supplies a constant current I_o and employs pnp transistors, whereas the sink removes a constant current I_o and uses npn devices. An essential feature of both circuits is that this current I_o is independent of external load impedance values. In both circuits the base of the output transistor Q_1 is clamped at a fixed potential determined by the base-to-emitter voltage V_{be} of the diode-connected biasing transistor Q_2. This diode connection stabilizes I_o against variations due to temperature because the base-emitter voltages are forced to track. For large transistor h_{fe} values the base currents can be neglected, and the emitter currents of these current-mirror circuits will be especially well matched when fabricated in integrated-circuit form. The impedance at the collector output terminal of these circuits is typically 1 M and is attributable to the small collector conductance element h_{oe}.

The application of the constant-current source as an active load Q_1 for the common-emitter amplifier Q_3 in Figure 2-6 creates an important linear electronic circuit. This technique is employed in almost all operational amplifiers to achieve their high open-loop gain A_{vo}. The current source is adjusted to the quiescent collector-current value of 6 mA by applying equation (2-4); simultaneously it presents an active load impedance $R_{o_1} = 1/h_{oe_1} = 1$ M. Remarkably, this would require an impractical 6-kV supply to achieve the same values with a passive load impedance. The now horizontal load line is very sensitive to variations in both the input signal and bias network, owing to the high voltage gain realized. Small changes in the amplifier base current produced by input signal swings result in symmetrical complementary output excursions in V_{ce} between Q_1 and Q_3. For bias stability the dual of the constant-current circuit, the constant-voltage configuration, is used for developing a stable voltage level in the presence of disturbing circuit influences. Transistor Q_4 provides temperature tracking and a path for the I_s variation of this common-emitter amplifier. Note that the emitter currents of Q_3 and Q_4 are essentially equal and that both bias resistors are determined from equation (2-4). Stability is further enhanced by the 100-Ω emitter resistor R_e on Q_3. This negative feedback increases the input impedance as described by equation (2-5). It also stabilizes the voltage gain by making it essentially equal to $-R_L/R_e$, which is independent of the transistor parameters. Voltage-gain determination at high collector load impedances must also include the parallel amplifier internal collector impedance $R_{o_3} = 1/h_{oe_3} = 1$ M, which compares with R_{o_1} and, together with it, establishes output symmetry between Q_1 and Q_3. The voltage gain is calculated by equation (2-7) and consists of the numerator of the common-emitter and denominator of the common-collector gain expressions from Table 2-1.

Transistors also contain an equivalent internal capacitance linking their input to output, shown as C_{cb} in Figure 2-6. The effect of this typical 4-pF capacitance in determining the frequency response of the common-emitter

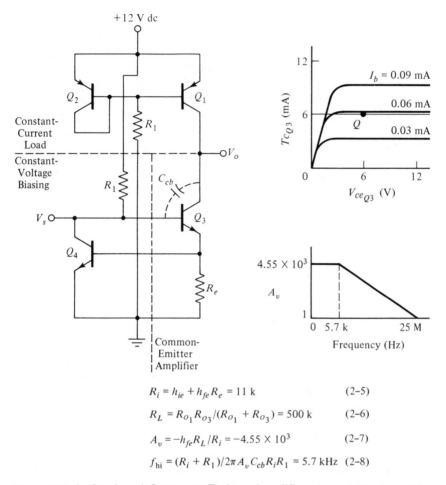

Figure 2-6 Active Load Common-Emitter Amplifier; $h_{fe} = 100$, $h_{ie} = 1\,k$, $R_0 = 1\,M$, $R_e = 100\,\Omega$, $R_1 = 1.8\,k$, $C_{cb} = 4\,pF$

$$R_i = h_{ie} + h_{fe}R_e = 11\ k \qquad (2\text{-}5)$$

$$R_L = R_{o_1}R_{o_3}/(R_{o_1} + R_{o_3}) = 500\ k \qquad (2\text{-}6)$$

$$A_v = -h_{fe}R_L/R_i = -4.55 \times 10^3 \qquad (2\text{-}7)$$

$$f_{hi} = (R_i + R_1)/2\pi A_v C_{cb}R_iR_1 = 5.7\ kHz \qquad (2\text{-}8)$$

amplifier is dominant at the base terminal, where C_{cb} is effectively multiplied by the amplifier voltage gain described by equation (2-8). The input time constant formed by the effective capacitance, resistor R_1, and input impedance R_i primarily determine amplifier cutoff f_{hi}. Even though the calculated cutoff frequency is 5.7 kHz, this amplifier stage does not reach unity gain until approximately 25 MHz because of the -20-dB/decade gain rolloff provided by its single-pole response. The minor effects of the output time constant on f_{hi} and other parallel impedances are ignored for clarity. The circuit of Figure 2-6 has a low input impedance and small deliverable output power. However, both can be improved by buffering the amplifier input and output with a common-collector stage. Therefore, common-collector stages generally provide input and output isolation, with common-emitter circuits serving as interior gain stages in cascaded amplifiers. An additional solution to the problem of interfacing linear circuits is frequently applied in operational amplifiers. This involves the use of base-current cancellation circuits to reduce input current by a factor of ten or more for bipolar devices, with an equivalent increase in input impedance. Also, collector neutralization circuits for the internal collector impedance element $1/h_{oe}$ can raise this impedance to the very high value of 10 M. Reference 2 has a further discussion of these techniques. Finally, if it is of interest to maximize both the gain and bandwidth of a common-emitter amplifier, rather than maximizing just the gain as in the example of Figure 2-6, a passive load impedance in the vicinity of 1 k is usually optimum.

2-3 DIFFERENTIAL AMPLIFIERS

The balanced dc differential amplifier shown in Figure 2-7(a) is a basic circuit used for a variety of linear applications, including the input stage of operational amplifiers. It is usually operated from symmetrical ± supplies applied in the normal polarities as shown for npn bipolar transistors with the base input terminals at 0 V under quiescent conditions. A positive-going signal applied to the base of either transistor will increase its collector current. Because of the interaction that results from this emitter-coupled arrangement, however, the effective drive for the differential amplifier is the algebraic difference of the input signals. Collector-current transfer curves are defined by equation (2-9) and Figure 2-7, where the abscissa represents normalized differential input voltage $(V_1 - V_2)/V_T$. It is apparent from the figure that for balanced input conditions $(V_1 - V_2 = 0)$ the emitter currents are equal to $\frac{1}{2} I_o$, and linear operation is available over a maximum differential input voltage swing of about ±25 mV at room temperature (±1 V_T unit). From Figure 2-7 it is implied that the transfer characteristics are a function of the volt-equivalent of temperature V_T, and the input-voltage-to-output-current transconductance is dependent upon the constant current I_o. Consequently, the versatility of this circuit can be expanded from that of an amplifier to an analog signal multiplier, if I_o is made a controllable variable multiplicand instead of constant current and the signal voltage the multiplier. An application of signal multiplication is presented in Figure 4-22.

The voltage gain $A_{v_{\text{diff}}}$ of the differential amplifier is expressed in terms of bipolar transistor h-parameters by equation (2-10), and the differential input impedance $R_{i_{\text{diff}}}$ by equation (2-11). Reexamination of Figure 2-7 shows Q_1 to be a common-collector amplifier and Q_2 a common-emitter amplifier with reference to output V_{o_2}. For an h_{fe} of 100 and h_{ie} of 1 k, the voltage gain of the common-collector stage is found to be 0.99999 from its formula in Table 2-1. This result, which is due to the swamping of h_{ie} by the high emitter load

Figure 2-7 Differential DC Amplifier and Normalized Transfer Curves; $h_{fe} = 100$, $h_{ie} = 1$ k

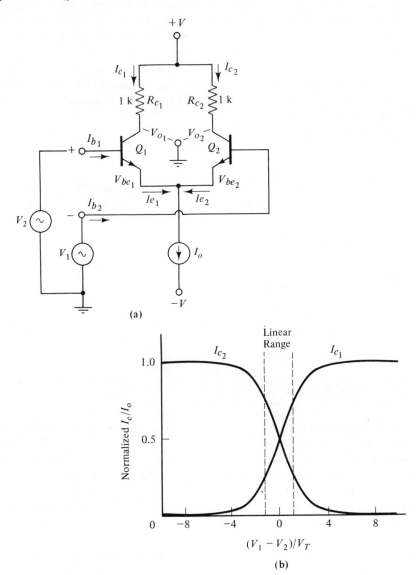

(a)

(b)

impedance R_o of 1 M provided by the constant-current sink, is responsible for the very nearly equal inverting and noninverting gains with reference to inputs V_2 and V_1. Consequently, connecting the bases and collectors of Q_1 and Q_2 together provides the common-mode connection that results in the very small voltage gain $A_{v_{cm}}$ expressed by equation (2-12) and the very high input impedance $R_{i_{cm}}$ of equation (2-13). The foregoing quantities are especially significant for differential amplifiers employed as the input stage in operational amplifiers. An important further relationship available from these gain quantities is the voltage common-mode rejection ratio (CMRR) defined by equation (2-14), which describes the ability of the amplifier to suppress common-mode input signals relative to differential input signals. This relationship is considered in greater detail in Chapter 3. Of equal concern, however, are the uncertainties produced by device temperature variation considered next.

$$I_o = \frac{h_{fe}}{(h_{fe} - 1)} (I_{c_1} + I_{c_2}) \qquad (2\text{-}9)$$

$$= I_{s_1} \cdot \exp (V_{be_1} / V_{T_1}) + I_{s_2} \exp (V_{be_2} / V_{T_2})$$

$$= 1 \text{ mA} \quad \text{typical}$$

$$A_{v_{\text{diff}}} = \frac{h_{fe} R_C}{2h_{ie}} = 50 \qquad (2\text{-}10)$$

$$R_{i_{\text{diff}}} = \frac{4V_T h_{fe}}{I_o} = 10 \text{ k} \qquad (2\text{-}11)$$

$$A_{v_{cm}} = \frac{1}{2} \frac{R_c}{R_o} = 5 \times 10^{-4} \qquad (2\text{-}12)$$

$$R_{i_{cm}} = h_{fe} R_o = 100 \text{ M} \qquad (2\text{-}13)$$

$$\text{CMRR} = \frac{A_{v_{\text{diff}}}}{A_{v_{cm}}} = 2 \times 10^5 \qquad (2\text{-}14)$$

The base-emitter voltages of a random group of the same type of bipolar transistors at the same collector current are typically within 20 mV of one another. In practice, operation of a differential pair from a constant-current sink permits a V_{be} match to within 1 mV. Equation (2-15) describes this input offset voltage V_{os}, showing it to depend primarily upon the mismatch in reverse saturation current I_s between the input transistors. This mismatch is due to variations in doping and geometry of the devices during their manufacture. Of greater concern, however, is the offset voltage drift with temperature dV_{os}/dT shown by equation (2-16). This input error, a difficult one to compensate, results from mistracking of the -2.5 mV/°C of V_{be_1} and

V_{be_2}. In the differential configuration dV_{os}/dT is reduced by a factor of about $\frac{1}{700}$ to 3.5 $\mu V/°C$ from the preceding base-emitter voltage match. Consequently, it should be appreciated that the differential amplifier connection primarily offers good dc stability.

Typical input bias-current offset and offset current drift are described by equations (2-17) and (2-18), where the temperature dependence of offset current averages $-0.5\%/°C$ and is attributable to a mismatch in current gains ($h_{fe_1} \neq h_{fe_2}$). These offset drift parameters, which constitute the principal input uncertainty of differential amplifiers, are of particular interest in the application of differential-input instrumentation amplifiers developed in Chapter 3. A final parameter of interest is the power-supply rejection ratio (PSRR) of equation (2-19). Since the differential amplifier is capable of amplifying dc voltages, it is inherently sensitive to changes in its ± power-supply voltage from imperfect regulation and ripple. This error appears at the input as a change in offset voltage V_{os} and, like the other voltage errors referred to the input, appears at the amplifier output multiplied by the voltage gain. PSRR is typically in the range of -80 dB, however, owing to the symmetry provided by the differential amplifier operating from balanced ± supplies.

$$V_{os} = V_{be_1} - V_{be_2} \tag{2-15}$$

$$= V_T \ell n \frac{I_{s_2}}{I_{s_1}} \cdot \frac{I_{e_1}}{I_{e_2}}$$

$$= 1 \quad mV$$

$$\frac{dV_{os}}{dT} = \frac{dV_{be_1}}{dT} - \frac{dV_{be_2}}{dT} \tag{2-16}$$

$$= 3.5 \quad \mu V/°C$$

$$I_{os} = I_{b_1} - I_{b_2} \tag{2-17}$$

$$= \frac{1}{2} \frac{(h_{fe_1} - h_{fe_2})I_o}{h_{fe_1} \cdot h_{fe_2}}$$

$$= 50 \quad nA$$

$$\frac{dI_{os}}{dT} = B \cdot I_{os} \qquad \begin{array}{l} B = -0.005/°C > 25°C \\ = -0.015/°C < 25°C \end{array} \tag{2-18}$$

$$= -0.25 \quad nA/°C$$

$$PSRR = \frac{dV_{os}}{dV_{supply}} \tag{2-19}$$

$$= 100 \quad \mu V/V$$

2-4 OPERATIONAL AMPLIFIERS

Operational amplifiers, so named by designers of analog computers, were a specialized gain block in early computer systems. During their development tubes, discrete transistors, and integrated-circuit active devices have been employed. The earliest integrated-circuit amplifier was offered in 1963 by Texas Instruments, but the Fairchild 709 introduced in 1965 was the first operational amplifier to achieve widespread application. Improvements in integrated-circuit design resulted in second-generation devices such as the National LM108. Then advances in fabrication technology, permitting mixed devices on a single substrate, led to third-generation devices, represented by the RCA 3140 amplifier, with improved specifications overall. The 1-volt-supply micropower operational amplifier, such as National's LM10, introduces fourth-generation devices that permit applications beyond conventional operational-amplifier practice, because they can be powered from natural or existing system potentials. Operational amplifiers are characterized by very high gain at dc and a uniform rolloff in this gain with frequency over many decades. This characteristic gives operational amplifiers their versatility, with the ability to accept feedback from a variety of networks with excellent dynamic stability. Consequently, the networks can accurately impart their parameters to circuit performance with negligible interaction or errors.

The equivalent circuit of an inverting operational amplifier is shown in Figure 2-8. Almost all operational amplifiers have a differential input and single-ended output so that the gain between V_o and V_1 is positive (noninverting), whereas the gain from V_2 to V_o is negative (inverting). For the ideal device the differential input impedance R_{i_diff} and open-loop voltage gain A_{v_o} approach infinity, and the output impedance R_o approaches zero. In practice, feedback is introduced between the output and inverting input, and a virtual ground is implied such that feedback from the output through R_f serves to maintain $V_d = V_1 - V_2$ at zero. No current actually flows into this implied short circuit, but it does provide a convenient model from which to determine the output voltage for an arbitrary input signal.

Other elemental operational-amplifier circuits are shown in Figure 2-9 with their respective voltage gains conveniently expressed in terms of resistor ratios. The accuracy of these simple gain equations is graphically described by the curves of Figure 2-10 in terms of device open-loop gain A_{v_o} and differential input impedance R_{i_diff}. For example, the maximum recommended closed-loop gain and input resistor values for the 741-type operational amplifier are found to be 1000 and 20 k, respectively. Gain linearity over the dynamic range of the amplifier is typically 0.01% referenced to the amplifier full-scale output, and is attributable to minor device nonlinearities such as associated with the input differential-pair constant-current sink. For the subtractor circuit a balancing resistor R_G must be connected to the input terminal, so that resistor ratios

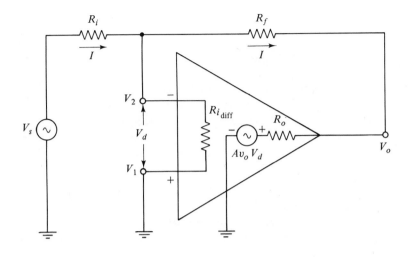

Since $Ri_{\text{diff}} \to \infty$ then $V_d = \dfrac{V_o}{Av_o} \to 0$ as $|Av_o| \to \infty$

$$\frac{V_o}{V_s} = \frac{-IR_f}{IR_i} = -\frac{R_f}{R_i}$$

(2-20)

Figure 2-8 Inverting Operational Amplifier

between the inverting and noninverting inputs are equal, as shown in Figure 2-9.

Most operational amplifiers are similar internally, as depicted by Figure 2-11, and include a differential-input stage, a high-gain innerstage employing the active-load technique, and a power-output stage. A primary utility of the operational amplifier is its ability to amplify dc signals stably and ac signals simultaneously without phase shift because of the direct interstage coupling. The availability of the inverting and noninverting inputs is of particular interest for instrumentation applications because of the common-mode signal-rejection capability it affords. However, we have already seen that the primary purpose of the differential-input stage that provides this rejection is to ensure good dc stability. The classification of operational amplifiers is typically determined by the active devices that implement the differential-input stage, since this stage principally determines the errors exhibited by the amplifier. Table 2-2 delineates this classification.

Buffer

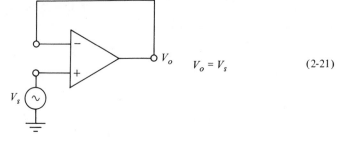

$$V_o = V_s \qquad (2\text{-}21)$$

Noninverter

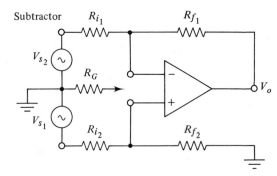

$$V_o = \left(1 + \frac{R_f}{R_1}\right) V_s \qquad (2\text{-}22)$$

$(2\text{-}23)$

Subtractor

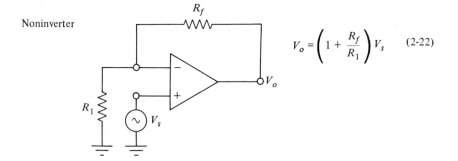

$$V_o = \frac{R_{f_2}}{R_{i_2}} \cdot V_{s_1} - \frac{R_{f_1}}{R_{i_1}} \cdot V_{s_2}$$

Connect R_G for balance:

$$\frac{R_{f_1}}{R_G} + \frac{R_{f_1}}{R_{i_1}} = \frac{R_{f_2}}{R_G} + \frac{R_{f_2}}{R_{i_2}}$$

Comparator

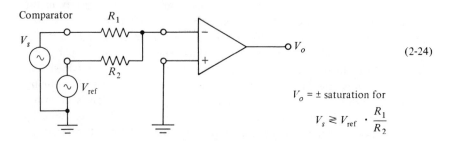

$(2\text{-}24)$

$V_o = \pm$ saturation for

$$V_s \gtrless V_{\text{ref}} \cdot \frac{R_1}{R_2}$$

Figure 2-9 Basic Operational-Amplifier Circuits

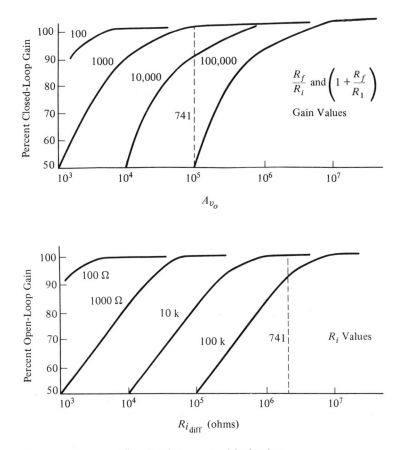

Figure 2-10 Amplifier Performance Limitations

Bipolar operational amplifiers require an input bias current I_b to maintain linear circuit operation. This current produces a voltage drop in the input resistors, which, when multiplied by amplifier gain, appears at the output as a voltage error. The input resistor of Figure 2-12 may be balanced by a compensating resistor R_c, with the result that the input bias-voltage drops become equal and are rejected by the CMRR of the amplifier. However, even with matched input resistors a residual error will remain, owing to the current-gain mismatch of the differential input-stage transistors. If this offset bias-current-produced voltage presents a problem, such as in a low-frequency active-filter application using large resistor values, selection of a better-grade operational amplifier is indicated. Input offset voltage V_{os}, whose cause was described in the previous section and by equation (2-15), when multiplied by the amplifier gain also appears at the output as a voltage error. The

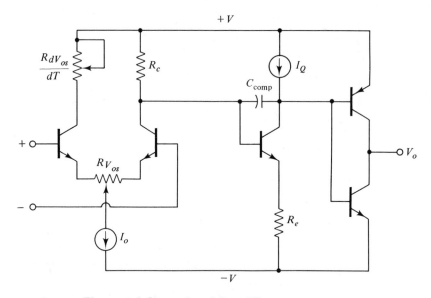

Figure 2-11 Elemental Operational Amplifier

manufacturer usually provides for a V_{os} adjustment by introduction of an external trimpot in the differential-input-stage emitter circuit, as shown in Figure 2-11. This permits the incremental addition and subtraction of emitter voltage, thereby zeroing V_{os} without disturbing the emitter current I_o.

Introducing an external trimpot into the tightly matched differential pair is not the best solution to offset adjustment because of the possible temperature drift of the trimpot resistance. Figure 2-13 shows preferred external offset-adjustment methods for the inverter and subtractor circuits, which are not sensitive to the trimpot temperature coefficient. The buffer amplifier is

Figure 2-12 Input Bias-Current Compensation

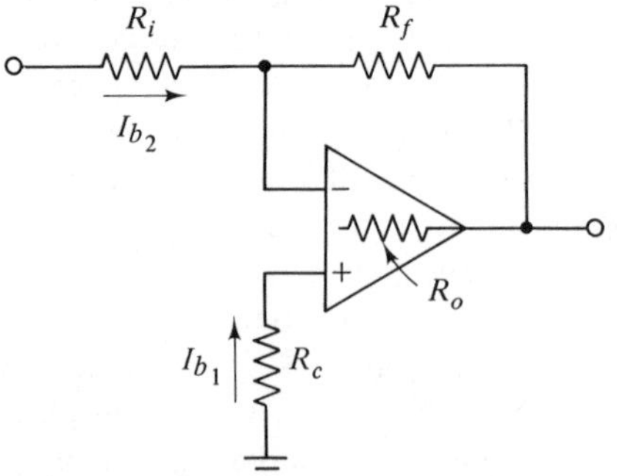

$$R_c = R_i \left(\frac{R_f + R_o}{R_i + R_f + R_o} \right) \text{ ohms} \qquad (2\text{-}25)$$

TABLE 2-2
OPERATIONAL AMPLIFIER TYPES

Bipolar	Prevalent type used for a wide range of signal-processing applications. Good balance of performance characteristics.
FET	Very high input impedance frequently employed as an instrumentation-amplifier preamplifier. Exhibits larger input errors than bipolar devices.
CAZ	Bipolar device with auto-zero circuitry for internally measuring and correcting input error voltages. Provides low-input-uncertainty amplification.
BiFET	Combined bipolar and FET circuit for extended performance. Intended to displace bipolar devices in general-purpose applications.
Superbeta	A bipolar device approaching FET input impedance with the lower bipolar errors. A disadvantage is lack of device ruggedness.
Micropower	High-performance operation down to 1-volt-supply powered from residual system potentials. Employs complicated low-power circuit equivalents for implementation.
Isolation	An internal barrier device using modulation or optical methods for very high isolation. Medical and industrial applications.
Chopper	DC-ac-dc circuit with a capacitor-coupled internal amplifier providing very low offset errors for minimum input uncertainty.
Varactor	Varactor diode input device with very low input bias currents for current-amplification applications such as photomultipliers.
Vibrating capacitor	A special input circuit arrangement requiring ultralow input bias currents for applications such as electrometers.

required for the subtractor circuit to provide a virtual ground for R_{f_2}. Use of a collector rheostat in Figure 2-11 instead of an emitter potentiometer will unbalance emitter current on one side of the differential pair, permitting the minimization of dV_{os}/dT through an imbalance in equation (2-15). Consequently, a 10% emitter-current imbalance will effect an 8-μV/$^{\circ}$C change in dV_{os}/dT.

2-5 GAIN AND BANDWIDTH RELATIONSHIPS

Operational amplifier performance at higher frequencies is represented by the Bode diagram of Figure 2-14, where the open-loop gain of the intrinsic device is denoted by the dashed straight-line approximation. This operational

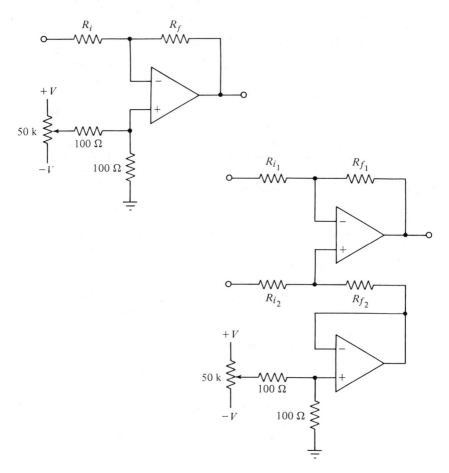

Figure 2-13 Preferred Offset Trim Methods

amplifier has three corner frequencies as shown, the lowest around 10 Hz, owing to the capacitor-compensated high-gain innerstage shown in Figure 2-11. The second corner frequency is in the vicinity of 1 MHz and is associated with the differential input stage. The third is perhaps at 25 MHz and is determined by the output stage. This composite gain-bandwidth characteristic has little practical application, however, until negative feedback is established between the output and inverting input, providing the closed-loop gain described by the straight-line projections. The difference between the open- and closed-loop gains is the loop gain, which is directly proportional to the dc stability, bandwidth, and noise and distortion reduction realized in practice. These relationships are described by equations (2-26) through (2-29).

According to negative-feedback theory, an inverting amplifier will be unstable if its gain is greater or equal to unity when the phase shift reaches 180° establishing an in-phase output-to-input relationship. Each corner frequency contributes 90° of phase shift, 45° at the corner. Consequently, the closure between the open-loop curve and closed-loop gain projection cannot proceed significantly beyond the second, or −40-dB/decade corner (135°), for unconditional stability. Higher closed-loop gains are more stable than lower closed-loop gains, in general, because the higher intercept entails less cumulative phase shift. However, caution should always be exercised in the amplification of signals in the region of diminishing loop gain (the difference between open- and closed-loop gains) because of the loss of its stabilizing influence.

It is ordinarily necessary to shape the gain-phase characteristic of operational amplifiers in order to ensure unconditional stability. The most

Figure 2-14 Amplifier Gain, Bandwidth, and Phase

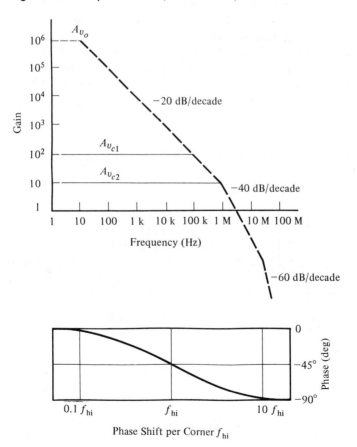

Phase Shift per Corner f_{hi}

common procedure is to synthesize a rolloff characteristic of -20 dB/decade extending approximately to the amplifier unity-gain axis as shown in Figure 2-14, which will be stable under all operating conditions. This dominant-pole phase-compensation technique is achieved through equation (2-8) by adding additional capacitance C_{comp} in Figure 2-11 at the high-gain innerstage, reducing the cutoff frequency to the illustrated 10 Hz. An alternate technique is pole-cancellation compensation, which uses external RC components to tailor the gain-phase characteristic to the closed-loop gain of interest for stable operation. The advantage of this latter technique is the extended bandwidth possible at higher closed-loop gains, realizable by following the compensation recommendations of the device manufacturer for the choice and connection of the RC elements.

In order to examine operational amplifier gain-bandwidth interactions, consider the example of two inverting amplifiers in cascade, each with a closed-loop gain $A_{v_{cl}}$ of 100 and 100 kHz bandwidth with reference to Figure 2-14. Consider also that each device is contributing 1% total harmonic distortion. We characterize this circuit by Figure 2-15 with the aid of equation (2-28), observing that distortion is additive. If it is required to reduce distortion to 0.1% per stage, then application of equation (2-27) indicates that closed-loop gain per stage $A_{v_{c2}}$ is reduced to 10. This modification is described by Figure 2-16, where for computation purposes the closed-loop gains of Figure 2-15 are taken as the open-loop gain entries in equations (2-26) and (2-27) for Figure 2-16.

Figure 2-15 Gain-Bandwidth Relationships

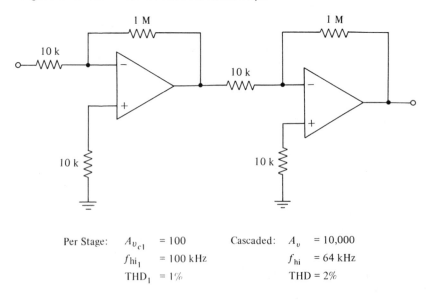

Per Stage:		Cascaded:	
$A_{v_{cl}}$	$= 100$	A_v	$= 10,000$
f_{hi_1}	$= 100$ kHz	f_{hi}	$= 64$ kHz
THD_1	$= 1\%$	THD	$= 2\%$

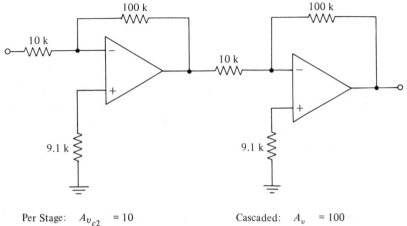

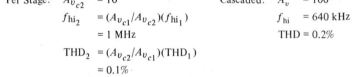

Per Stage: $A_{v_{c2}}$ = 10   Cascaded: A_v = 100

f_{hi_2} = $(A_{v_{c1}}/A_{v_{c2}})(f_{hi_1})$ f_{hi} = 640 kHz

        = 1 MHz   THD = 0.2%

THD_2 = $(A_{v_{c2}}/A_{v_{c1}})(THD_1)$

        = 0.1%

Figure 2-16 Gain-Bandwidth Modification

f_{hi} and dc stability multiplier $= \dfrac{A_{v_o}}{A_{v_c}}$ *(2-26)*

distortion and noise reduction $= \dfrac{A_{v_c}}{A_{v_o}}$ *(2-27)*

cascaded-stages bandwidth $= \dfrac{0.35}{1.1[(0.35/f_{hi_1})^2 + \cdots + (0.35/f_{hi_N})^2]^{1/2}}$ *(2-28)*

voltage gain of cascaded stages $= A_{v_{c1}} \cdot A_{v_{c2}} \cdots A_{v_{cN}}$ *(2-29)*

A concern is how to maximize the bandwidth for a given voltage-gain requirement in a cascade of stages. Assuming a conventional single-pole amplifier rolloff as in Figure 2-14, obtaining a gain significantly greater than unity with a single stage provides a diminishing bandwidth. It is developed in Reference 10 that a cascade of identical stages each with a voltage gain of $e^{\frac{1}{2}} = 1.65$ provides the maximum amplifier bandwidth, but this is too inefficient because of the number of stages required. Figure 2-17 indicates that the minimum number of identical stages that is optimum in increasing the overall bandwidth occurs at $e = 2.718$ stages. The overall bandwidth for a cascade of n identical amplifiers each of gain A_{v_c} and high-frequency cutoff f_{hi}

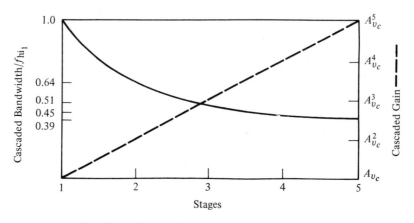

Figure 2-17 Relationship for Cascaded Identical Stages

varies as $\sqrt{2^{1/n} - 1} \cdot f_{\mathrm{hi}}$, and for three stages it is $0.51 \cdot f_{\mathrm{hi}}$. It is also occasionally necessary to evaluate settling time for a single-pole RC response. This need arises, for example, in determining the number of time constants

Figure 2-18 RC Time-Constant Amplitude Error

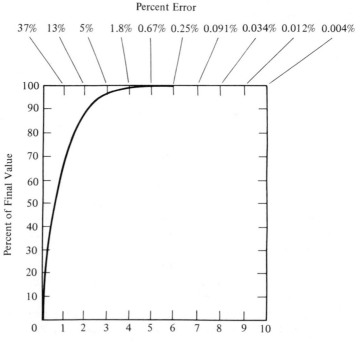

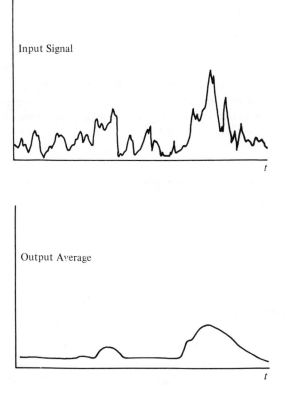

Figure 2-19 Signal Averaging

required for an amplifier or signal-acquisition circuit to settle to the accuracy of interest. Figure 2-18 provides the amplitude error as a function of the number of elapsed time constants.

2-6 *SIGNAL AVERAGING*

When signal and interference frequencies are coincident such that the signal and noise bandwidths are the same, linear filtering cannot separate the two to improve signal quality. A technique useful in this case, if the signal is repetitive, is signal averaging. We can understand this operation by considering that if a periodic signal is averaged for n signal periods of duration T, then the output will be composed of arithmetic signal and rms interference additions. For example, after averaging 100 periods the output is improved by a factor of $100/\sqrt{100}$, or tenfold. Improvement in signal quality due to averaging is illustrated by Figure 2-19.

A practical realization for determining averages of signals having rapid fluctuations about a mean value that also may vary—that is, a nonstationary average value—is described by the finite time-averaging process of equation (2-30). The noninverting Paynter running-average mechanization of Figure 2-20 is a stable third-order circuit having a response continuously approximating an ideal average over the period of the signal of interest. The RC time constant of this circuit is chosen to equal $T/2\pi$, where T is the signal period to be averaged. The network then sharply attenuates all signals having periods shorter than T (higher frequencies). For very weak and noisy instrumentation signals a lock-in amplifier, such as available from Princeton Applied Research, Incorporated, may be employed to obtain usable signal levels. This higher-performance signal averager uses more complicated modulation and narrow-bandwidth methods and is capable of improvement to 10^{10}.

$$V_o = \frac{1}{T} \int_{t-T}^{t} V_i \cdot dt \qquad (2\text{-}30)$$
$$= \overline{V_i}$$

2-7 LOGARITHMIC OPERATIONS

The utility of logarithmic functions in signal processing primarily lies with their ability to accommodate wide-dynamic-range signals. Applications include signal compression and expansion, signal-to-noise ratio enhancement, and the

Figure 2-20 Paynter Running-Average Circuit (*With permission of Teledyne-Philbrick*)

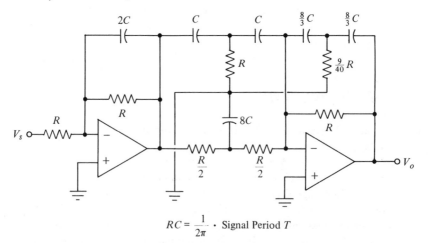

$$RC = \frac{1}{2\pi} \cdot \text{Signal Period } T$$

TABLE 2-3
LOGARITHMIC FUNCTIONS

FUNCTION	DESCRIPTION
$V_o = C \cdot \log_{10}\left(\dfrac{V_i}{V_{ref}}\right)$	Voltage log
$V_o = C \cdot \log_{10}\left(\dfrac{I_i}{I_{ref}}\right)$	Current log
$V_o = V_{ref} \cdot 10^{-V_i/C}$	Antilog
$V_o = C \cdot \log_{10}\left(\dfrac{V_1}{V_2}\right)$	Log ratio
$V_o = C \cdot \sinh^{-1}\left(\dfrac{V_i}{2V_{ref}}\right)$	Bipolar log
$V_o = C \cdot \sinh\left(\dfrac{V_i}{V_{ref}}\right)$	Bipolar antilog

mechanization of computations. Logarithmic arguments are always dimension-less. Consequently, the log of the ratio of two voltages or currents is required in practice, with the denominator normally a fixed reference value as shown in Table 2-3. Commercially available log devices offer performance extending over four voltage decades, typically 1 mV to 10 V, and six current decades, from 1 nA to 1 mA. The log ratio device is useful over seven decades of signal amplitude, with four decades in the numerator and three in the denominator. The bipolar log function is symmetrical about and linear in the vicinity of zero input voltage, as illustrated by Figure 2-21. The mathematical representation of this function is the inverse hyperbolic sine ($\sinh^{-1}$), typically extending over ± 3 decades, implemented by Euler's identity with a pair of complementary antilog exponential functions. Equation (2-31) describes this function.

Present logarithmic circuit realizations employ silicon bipolar devices in matched monolithic pairs for good thermal stability. Troublesome bias-current errors are usually minimized by employing FET-input operational amplifiers. The basic logarithmic relationship is derived from the diode equation of equation (2-32), where $V_o = V_f$ and $I_f = V_i/R$ with reference to the circuit of Figure 2-22. Input voltage or current for the log devices is positive and the output negative for $V_i > V_{ref}$ or $I_i > I_{ref}$, and reversed for $V_i < V_{ref}$ or $I_i < I_{ref}$. The bandwidth for commercially available log devices changes with the

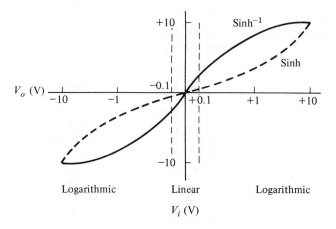

Figure 2-21 Bipolar Log and Antilog Functions

input signal level, varying between 100 Hz and 100 kHz for minimum to maximum signal levels, respectively. Log conformity accuracy typically is within 1% throughout the dynamic range.

$$V_i = V_{ref} \; 10^{V_o/C} - V_{ref} \; 10^{-V_o/C} \qquad\qquad (2\text{-}31)$$

$$\frac{V_i}{2V_{ref}} = \frac{\exp\,(2.3V_o/C) - \exp\,(-2.3V_o/C)}{2}$$

$$= \sinh\left(\frac{2.3V_o}{C}\right)$$

Rearranging, $$V_o = C \sinh^{-1}\left(\frac{V_i}{2V_{ref}}\right)$$

Figure 2-22 Basic Logarithmic Circuit

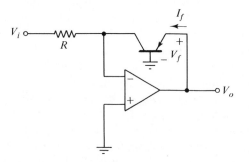

$$I_f = I_s \left[\exp \left(\frac{V_f}{V_T} \right) - 1 \right] \tag{2-32}$$

$$\ln I_f = \ln I_s + \frac{V_f}{V_T}$$

Rearranging, $\quad V_o = V_T (\ln I_f - \ln I_s)$

$$= \ln \left(\frac{V_i}{R} \right) - \ln I_s$$

$$= V_T \ln \left(\frac{V_i}{RI_s} \right)$$

$$= C \, \log_{10} \left(\frac{V_i}{V_{\text{ref}}} \right)$$

The application of a logarithmic function at the input of a data-conversion system allows reduction of the A/D-converter wordlength. This technique has merit for processing high-resolution sensor data by truncated-wordlength processors. Input compression increases the number of A/D quantizing levels at lower signal amplitudes, producing an effective decrease in quantization noise at these levels. Following conversion and digital computation, however, an antilog operation is required to restore signal linearity. The data may be antilogged by software, or by an analog antilog function following output signal reconstruction. Input compression followed by output expansion, or *companding*, also enhances signal-to-noise ratio, which may be of interest for other applications such as remote signal transmission. Companding is a nonlinear signal operation, however, which produces harmonics and requires a wider signal bandwidth than linear signals.

By way of example, a 13-bit-quality signal existing between 1 mV and 10 V can be represented by an eight-bit binary wordlength using logarithmic compression. A unipolar four-decade voltage-transfer characteristic is shown in Figure 2-23, where the scale factor C is adjusted for 2.5 V/decade. This permits a 0- to 10-V compressor output, when offsetting is provided as shown, for scaling to a 10-V full-scale A/D converter. It is apparent that the logarithmic gain is unity at the 10-V full-scale input level, increasing to a maximum at the minimum input level. A mirror-symmetry antilog operation is required following output-signal reconstruction for linear signal representation as shown in Figure 2-24. With the typical 1% log conformity error, an eight-bit A/D LSB of 39.1 mV more than satisfies the 1% of 10 V full-scale minimum value of 100 mV. Consequently, a constant fractional error is maintained throughout the signal dynamic range in logarithmic form at the expense of high resolution at any point within the range.

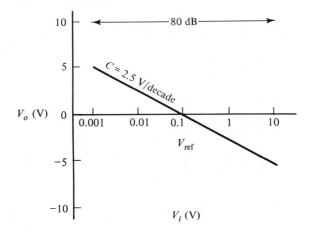

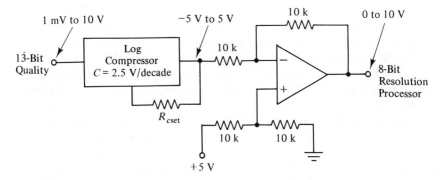

Figure 2-23 Logarithmic Compression

2-8 ANALOG COMPUTATION

The requirement occasionally arises in measurement and control systems to perform computations, such as an rms extraction or volumetric flow determination, in analog form. An especially versatile device for analog computation is the logarithmic-based multifunction module, which combines multiplication and division with the ability to raise a voltage or voltage ratio to an arbitrary positive or reciprocal power. The magnitude of the power may be greater than unity (power) or less than unity (root). These devices are available from several manufacturers including Analog Devices, Teledyne-Philbrick, and Intronics, with errors to 0.25% of their full-scale output. They typically implement an equation such as (2-33), in which the X, Y, Z variables can assume positive values between 0 and 10 V, and M may be any constant

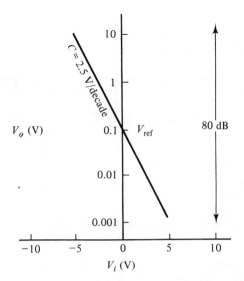

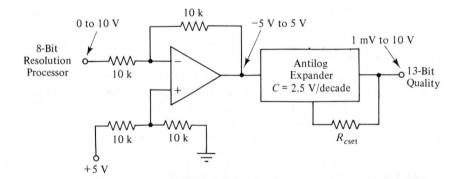

Figure 2-24 Antilog Expansion

between 0.2 and 5. This combination of mathematical functions within a single module is comparable to a small analog computer.

$$V_o = Y \cdot \left(\frac{Z}{X} \right)^M \qquad (2\text{-}33)$$

Mechanization of the log-antilog multifunction module is shown in block-diagram form in Figure 2-25. Its symmetrical circuit arrangement provides very low scale-factor and offset errors, and internally generated noise is of the order of 100 μV. Bandwidth is a function of the input level varying from about 1 kHz to 100 Hz for a 10-mV to 10-V input level, respectively.

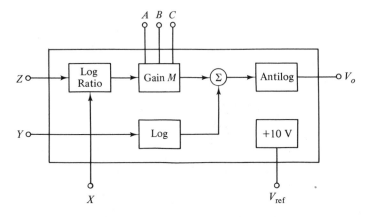

Figure 2-25 Log-Antilog Multifunction Module

The multifunction module can be connected to perform numerous functions directly plus additional ones with external circuitry, several of which are tabulated in Table 2-4. The circuit implementations of two examples are shown in Figures 2-26 and 2-27, where the latter provides volumetric and mass-flow computation of process data. The computations performed are described by equations (2-34) and (2-35) and Table 2-5, and consist of 15 mathematical operations. The rationale for an analog realization is the need for local reduction of process data involving multiple analog inputs, which presently can be provided at a cost below that for a microprocessor-based system with analog I/O in this range of mathematical operations. In the example presented, the flow-rate computer is for a flue-gas SO_2 scrubber on a fossil-fuel power plant, and the analog-computed flow data are transmitted to a remote supervisory-control digital computer.

TABLE 2-4
MULTIFUNCTION MODULE OPERATIONS

FUNCTION	TRANSFER EQUATION
Multiplication	$YZ/10$
Division	$10(Z/X)$
Root of ratio	$Y(Z/X)^M \quad M < 1$
Power of ratio	$Y(Z/X)^M \quad M > 1$
Reciprocal power	$Y(Z/X)^M = Y(X/Z)^{-M}$
True rms	$\overline{V_i^2}/V_o$
Trigonometric	$\sin Z = Z - 0.17(Z)^{2.83}$
	$\cos Z = 1 + 0.235Z - 0.7(Z)^{1.5}$

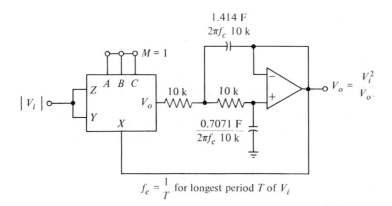

$$f_c = \frac{1}{T} \text{ for longest period } T \text{ of } V_i$$

Figure 2-26 True RMS Extraction

$$Q = A_n \cdot C_p \cdot K_p \cdot \sqrt{\frac{\Delta p \cdot T}{P \cdot M_w}} \cdot \frac{P \cdot T_{\text{std}}}{P_A \cdot T} \frac{m^3}{s} \qquad (2\text{-}34)$$

$$= (1 \text{ V}) (3.5) \cdot \left[\frac{(1.2 \text{ V}) (4.23 \text{ V})/10 \text{ V}}{(9.0 \text{ V}) (2.8 \text{ V})/10 \text{ V}} \right]^{1/2} \cdot \frac{(9.0 \text{ V}) (0.293)}{(7.2 \text{ V}) (4.23 \text{ V})/10 \text{ V}}$$

$$= 1.35 \text{ V corresponding to } 1350 \text{ m}^3/\text{s volumetric flow}$$

$$M = \frac{Q \cdot \text{ppm} \cdot M_{w_s}}{22.4} \frac{\text{kg}}{\text{s}} \qquad (2\text{-}35)$$

$$= \frac{(1.35 \text{ V}) (0.303 \text{ V}) (2.86)}{10 \text{ V}}$$

$$= 0.117 \text{ V corresponding to } 1.117 \text{ kg/s SO}_2 \text{ mass flow}$$

A frequent signal-processing requirement is to perform a precision ac-to-dc rectification for signals down to millivolt levels. Passive full-wave rectification is inadequate for this task, because silicon diodes will not conduct until the applied voltage reaches about 600 mV. However, the active full-wave rectification and smoothing circuit shown in Figure 2-28 can provide accurate rectification to submillivolt signal levels. About 99.4% conversion accuracy is realizable, determined primarily by the residual harmonic distortion in the output. For the signal-pole RC smoothing filter, a cutoff of $\omega_c = 1/RC$ radians placed one decade below the signal frequency ω will provide this accuracy. Equation (2-36) presents the first three Fourier terms for a fullwave rectified sinusoidal signal, and Table 2-6 the ac-to-dc converter response. This circuit will also provide the absolute value of the input signal with removal of the capacitor C.

$$V_o = \frac{2}{\pi} \cdot V_m + \frac{4V_m}{3\pi} \cdot \cos 2\omega t + \frac{4V_m}{15\pi} \cdot \cos 4\omega t \qquad (2\text{-}36)$$

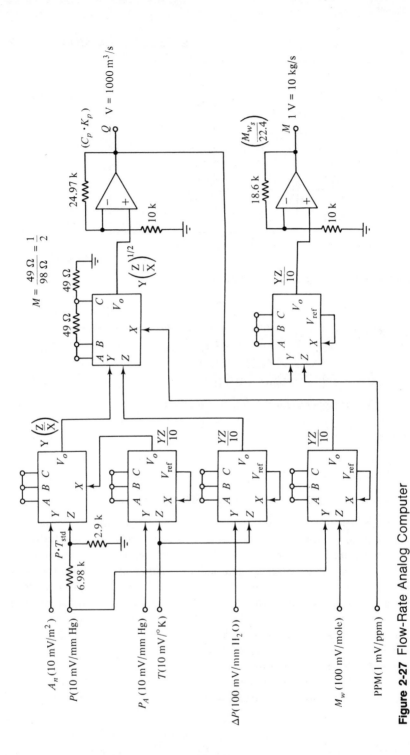

Figure 2-27 Flow-Rate Analog Computer

48

TABLE 2-5
FLOW COMPUTER INPUT PARAMETERS

PARAMETER	VALUE	SCALING	DESCRIPTION
A_n	$100\ m^2$	$10\ mV/m^2$	Stack area
P	$900\ mmHg$	$10\ mV/mmHg$	Stack absolute pressure
T_{std}	$293°K$	0.293 attn	Standard temperature
P_A	$720\ mmHg$	$10\ mV/mmHg$	Absolute atmospheric pressure
T	$423°K$	$10\ mV/°K$	Stack absolute temperature
ΔP	$12\ mmH_2O$	$100\ mV/mmH_2O$	Stack differential pressure
M_w	28 moles	$100\ mV/mole$	Total effluent molecular weight
ppm	$303\ SO_2$	$1\ mV/ppm$	Stack chemical analyzer output
M_{w_s}	64 moles SO_2	64 gain	Compound molecular weight
C_p	1.0	Unity	Sampling probe coefficient
K_p	$34.97\ m^2/s$	3.497 gain	Dimensional constant $\left[\dfrac{(g/g\ mole)(mmHg)}{(°K)(mmH_2O)}\right]^{1/2}$

TABLE 2-6
AC-DC COMPUTER RESPONSE

PARAMETER	V	V_o	% dc
DC component	$0.636V_m$	$0.636V_m$	99.4
Second harmonic	$0.424V_m$	$4.2 \times 10^{-3}V_m$	0.6
Fourth harmonic	$0.085V_m$	$8.5 \times 10^{-6}V_m$	1.3×10^{-3}

Figure 2-28 Precision AC-to-DC Converter

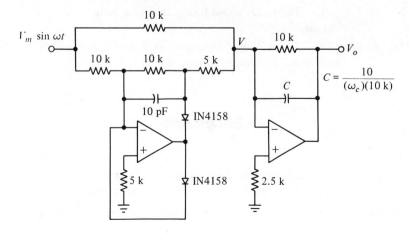

REFERENCES

1. J. MILLMAN, *Microelectronics*, McGraw-Hill, New York, 1979.
2. J. EMBINDER, *Application Considerations for Linear Integrated Circuits*, Wiley-Interscience, New York, 1970.
3. G. B. RUTKOWSKI, *Handbook of Integrated-Circuit Operational Amplifiers*, Prentice-Hall, Englewood Cliffs, N.J., 1975.
4. F. C. FITCHEN, *Electronic Integrated Circuits and Systems*, Van Nostrand Reinhold, New York, 1970.
5. RCA Linear Integrated Circuits, Technical Series IC-41, 1967.
6. G. TOBEY, J. GRAEME, and L. HUELSMAN, *Operational Amplifiers: Design and Applications*, McGraw-Hill, New York, 1971.
7. J. I. SMITH, *Modern Operational Amplifier Circuit Design*, John Wiley, New York, 1971.
8. J. A. CONNELLY, *Analog Integrated Circuits*, Wiley-Interscience, New York, 1975.
9. J. M. PETTIT and M. M. MCWHORTER, *Electronic Amplifier Circuits*, McGraw-Hill, New York, 1961.
10. THORNTON, SEARLE, PEDERSON, ADLER, and ANGELO, *Multistage Transistor Circuits*, SEEC Vol. 5, John Wiley, New York, 1965.
11. R. C. DOBKIN, "Logarithmic Converters," *IEEE Spectrum*, November 1969.
12. R. G. DURNAL, "Approximating Waveforms with Exponential Functions," *Electronics*, February 1, 1973.
13. R. KREAGER, "AC-to-DC Converters for Low-Level Input Signals," *Electronic Design News*, April 5, 1973.
14. D. R. MORGAN, "Get the Most Out of Log Amplifiers by Understanding the Error Sources," *Electronic Design News*, January 20, 1973.
15. G. NIU, "Get Wider Dynamic Range in a Log Amp," *Electronic Design*, February 15, 1973.
16. A. SANTONI, "True RMS Measurements Reveal the Power Behind the Waveform," *Electronics*, March 18, 1976.

PROBLEMS

2-1. A toy walkie-talkie includes the following CE speech amplifier. Find the voltage gain A_v including the input voltage-divider effect, and the high-frequency cutoff f_{hi}. Ignore the blocking capacitor.

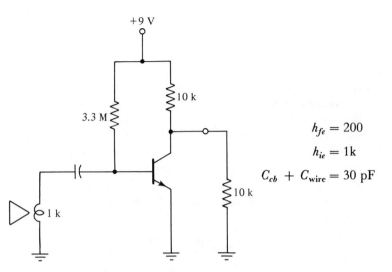

2-2.   Design an operational amplifier subtractor circuit by Figure 2-9 to convert a 1-mV/°C thermocouple input to a 1-mV/°K output. All signals are positive, 0°C equals 273°K, and all feedback resistors are to be 100 k.

2-3.   Determine the overall voltage gain A_v of the cascaded amplifier including the input voltage divider by working from the output to the input. Ignore biasing considerations. Assume $h_{fe} = 50, h_{ie} = 1.1$ k.

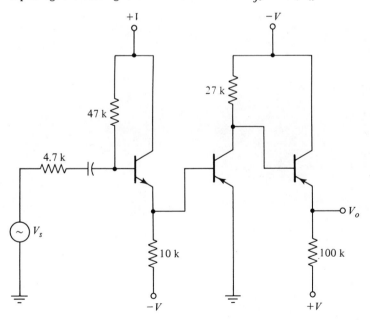

2-4.   A noisy periodic signal having a 100-Hz fundamental is to be upgraded through signal averaging. Design a Paynter running-average circuit for this signal using 10-k resistors.

2-5.   A four-decade (80-dB) dynamic range seismic signal is logarithmically compressed between 0.1 and 1 V-peak for recording purposes. Characterize a bipolar expander function from Table 10-1 to recover this signal with a 10-V_{FS} output. Tabulate the input and output signals to aid in fitting the expander function.

2-6.   The implicit use of an output variable in the solution of an analog computation as a result of feedback generally results in improved accuracy. Implement a circuit for the vector-sum identity below with a multifunction module and operational amplifiers, including all component values.

$$V_o = \sqrt{V_x^2 + V_y^2}$$
$$= V_y + \frac{V_x^2}{V_o + V_y}$$

instrumentation amplifiers

3-0 INTRODUCTION

Instrumentation amplifiers are usually the first electronic components encountered at the input of signal-acquisition and conditioning systems. Their primary purpose is to interface measurement signals to data-acquisition and conversion systems with accurate voltage amplification, and their difference from operational amplifiers is due principally to the common-mode signal rejection and ground-return potential difference attenuation provided plus a minimized input uncertainty error.

The first section of this chapter develops the evolution of instrumentation amplifiers and their selection; the discussion progresses from single-amplifier subtractor circuits and multiple-amplifier committed-gain circuits, through isolation instrumentation amplifiers for applications requiring source isolation, to amplifiers optimized for low input uncertainty or wideband service. The next section examines internal noise sources and their effect on low-level signal amplification including measurement error. A final section develops a detailed instrumentation-amplifier error analysis, which is shown to be essential for quantitatively determining amplifier performance and for identifying where error-reduction efforts are most productive.

3-1 INSTRUMENTATION-AMPLIFIER SELECTION

The accurate measurement of low-level signals in data-acquisition systems generally requires instrumentation amplifiers that can provide a large CMRR in order to reject interference sources coupled to sensor lines. Also essential is a high amplifier input impedance to preclude input signal loading effects from finite source impedances, and to accommodate source-impedance imbalances without degrading CMRR. An additional and important requirement is that instrumentation amplifiers possess sufficient linearity and stability to minimize parameter drift with time and temperature variations. This latter consideration results in the majority of the error terms that make up the amplifier error budget developed subsequently.

The relationship of CMRR to the output signal V_o of an instrumentation amplifier is described by equation (3-1), which is based on the definition of CMRR provided by equation (2-14), repeated here for convenience. For the single-amplifier subtractor instrumentation amplifier shown by Figure 3-1, $A_{v_{\text{diff}}}$ is determined by the ratios of $R_{f_1}/R_{i_1} = R_{f_2}/R_{i_2}$, and $A_{v_{cm}}$ by the mismatch in the R_i and R_f resistor tolerances. Consequently, its CMRR may be obtained from equation (3-2), where Table 3-1 presents a tabulation of the average CMRR for specific resistor tolerances with $R_{f_1} = R_{f_2} = R_{i_1} = R_{i_2}$. Expected values of CMRR at any $A_{v_{\text{diff}}}$ are then easily determined by gain scaling for specified resistor tolerances. CMRR is therefore seen to be a function of both $A_{v_{\text{diff}}}$ and the effect on $A_{v_{cm}}$ due to resistor tolerance.

Figure 3-1 Subtractor Instrumentation Amplifier

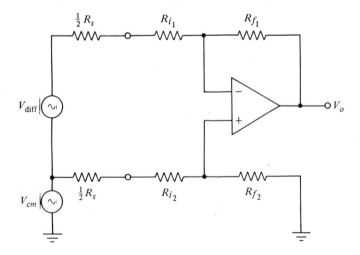

$$V_o = A_{v_{\text{diff}}} \cdot V_{\text{diff}} + A_{v_{cm}} \cdot V_{cm} \qquad (3\text{-}1)$$

$$= A_{v_{\text{diff}}} \cdot V_{\text{diff}} \left(1 + \frac{1}{\text{CMRR}} \cdot \frac{V_{cm}}{V_{\text{diff}}} \right)$$

$$\text{CMRR} = \frac{A_{v_{\text{diff}}}}{A_{v_{cm}}} \qquad (2\text{-}14)$$

$$\text{CMRR}_{\text{subtractor}} = \frac{\dfrac{1}{2}\left(\left| \dfrac{R_{f_2} \pm \Delta R_{f_2}}{R_{i_2} \pm \Delta R_{i_2}} \right| + \left| \dfrac{R_{f_1} \pm \Delta R_{f_1}}{R_{i_1} \pm \Delta R_{i_1}} \right| \right)}{\left| \dfrac{R_{f_2} \pm \Delta R_{f_2}}{R_{i_2} \pm \Delta R_{i_2}} \right| - \left| \dfrac{R_{f_1} \pm \Delta R_{f_1}}{R_{i_1} \pm \Delta R_{i_1}} \right|} \qquad (3\text{-}2)$$

$$\text{CMRR}_{\text{in-circuit}} = \frac{\text{CMRR}_{\text{subtractor}} \cdot \text{CMRR}_{\text{intrinsic}}}{\text{CMRR}_{\text{subtractor}} + \text{CMRR}_{\text{intrinsic}}} \qquad (3\text{-}3)$$

TABLE 3-1
CMRR VERSUS RESISTOR TOLERANCE

Resistor tolerance (%)	5	2	1	0.1
$A_{v_{cm}}$ subtractor (average)	0.1	0.04	0.02	0.002
$\text{CMRR}_{\text{subtractor}}$ $(A_{v_{\text{diff}}} = 1)$	10	25	50	500

Equation (3-3) defines the actual in-circuit CMRR, which is the parallel combination of intrinsic CMRR specified by the manufacturer and that determined by the external resistor values and their tolerances.

The single-amplifier instrumentation amplifier is capable of useful values of CMRR, but its realization requires the matching of four resistors and has an input impedance limited to the value of R_i. For bipolar amplifiers the typical value of R_i is in the 10-k range, and this can result in signal-loading effects for large values of source impedance R_s associated with V_{diff}. An additional problem is the effect of an imbalance in these source impedances in reducing CMRR. For high common-mode rejection the allowable mismatch is very small. For example, from equation (3-2) achieving a CMRR of 1000 to 1 at unity $A_{v_{\text{diff}}}$ allows an overall source impedance mismatch of one part per thousand, or 10 ohms for an R_i of 10 k. An improvement in this source-impedance constraint is available with the use of JFET-input operational amplifiers, with which R_i values to 1 M are practical. BiFET single-amplifier instrumentation amplifiers, such as the Analog Devices AD542, consequently enjoy widespread and economical application with realizable CMRR values to 10^4. BiFET devices have displaced earlier FET-input amplifiers, both of

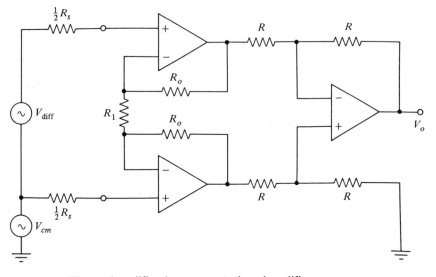

Figure 3-2 Three-Amplifier Instrumentation Amplifier

which have bipolar gain and output stages. The essential difference is that BiFET amplifiers employ ion implantation to improve matching of the JFET input devices, and they are superior to trimmed FET-input amplifiers with their dV_{os}/dT compromises.

The three-amplifier instrumentation amplifier of Figure 3-2 offers the advantage of very high input impedance with up to 1 k source-impedance imbalance without derogation of CMRR. This results from the high-impedance noninverting-amplifier inputs, which is the standard input arrangement for most instrumentation amplifiers. Each input impedance R_i is derived by equation (3-7) in terms of the differential and common-mode differential amplifier input impedances defined in Chapter 2, and the instrumentation amplifier open- and closed-loop gains. For present-day devices A_{vo} (open-loop gain) is 10^6, and input impedance is essentially determined by $R_{i_{cm}}$. Of interest is the dual-amplifier input stage, which has unity $A_{v_{cm}}$ owing to its differential input and differential output capability. As a result, equation (3-4) is an expression for both the first-stage $A_{v_{diff}}$ and its CMRR. In order to minimize the amplification of noise and offset voltage from the first stage, the $A_{v_{diff}}$ of the subtractor circuit is normally set at unity and all of the differential voltage gain is obtained with the input section. The $CMRR_{3ampl}$ of this arrangement is therefore the product of the $A_{v_{diff}}$ and the single-amplifier CMRR tabulated in Table 3-1 for the specific resistor tolerance chosen. This is described by equation (3-5). Three-amplifier instrumentation amplifiers are available in monolithic and hybrid form and include resistor tolerances to 0.1% or better, input impedances R_i to 1 gigohm (10^9 ohms), differential voltage gains to 10^3 settable by a single resistor, and CMRR values to 10^5 at $A_{v_{\text{diff}}}$ values of 100 and above. This three-amplifier version, whether in integrated form or constructed from discrete devices, offers a factor of ten or greater CMRR and allowable source-impedance imbalance in comparison with the single-amplifier instrumentation amplifier. The Analog Devices AD 522 is an example of a device with these performance capabilities.

$$\text{CMRR}_{\text{1st stage}} = A_{v_{\text{diff}}} \qquad\qquad (3\text{-}4)$$

$$= 1 + \frac{2R_o}{R_1}$$

$$\text{CMRR}_{3\ \text{ampl}} = \text{CMRR}_{\text{1st stage}} \cdot \text{CMRR}_{\text{subtractor}} \qquad (3\text{-}5)$$

$$\text{CMRR}_{\text{in-circuit}} = \frac{\text{CMRR}_{\text{1st stage}} \cdot \text{CMRR}_{\text{intrinsic}}}{\text{CMRR}_{\text{1st stage}} + \text{CMRR}_{\text{intrinsic}}} \qquad (3\text{-}6)$$

$$\cdot \frac{\text{CMRR}_{\text{subtractor}} \cdot \text{CMRR}_{\text{intrinsic}}}{\text{CMRR}_{\text{subtractor}} + \text{CMRR}_{\text{intrinsic}}}$$

$$R_i = \frac{1}{\dfrac{A_{v_{\text{diff}}}}{A_{v_o} R_{i_{\text{diff}}}} + \dfrac{1}{R_{i_{cm}}}} \qquad (3\text{-}7)$$

High-performance instrumentation amplifiers typically offer an additional factor-of-ten improvement in key performance parameters over the three-amplifier version. High-performance amplifiers are available in modular or hybrid circuit form because of the necessity for the manufacturer to trim the internal resistor networks. Their circuit generally is an extension of the three-amplifier configuration with the addition of the preamp, plus input-stage common-mode signal sensing and feedback to the preamp current sources for improved CMRR. This circuit, illustrated by Figure 3-3, can provide CMRR values to 10^6 at an $A_{v_{\text{diff}}}$ of 10^3, the latter expressed by equation (3-8). The choice of a high-performance instrumentation amplifier, such as the hybrid National LH 0038, is as likely to be based on the need for improved stability and reduced parameter drift with time and temperature as the need for higher CMRR. A comparison of all these specifications is provided in Table 3-2 (page 64) for the instrumentation-amplifier types introduced in this section.

$$A_{v_{\text{diff}}} = 1 + \frac{2R_o}{R_1} \qquad\qquad (3\text{-}8)$$

Many applications require a very low amplifier input uncertainty, such as precision low-level signal measurement over an extended time period in

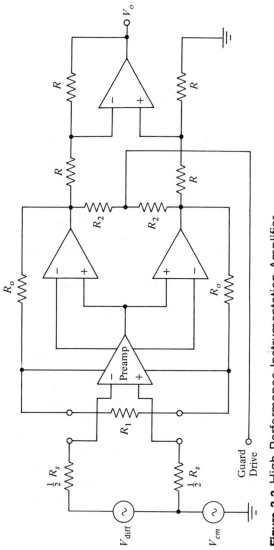

Figure 3-3 High-Performance Instrumentation Amplifier

which the amplifier may be subjected to temperature variations. With reference to the input errors associated with differential amplifiers presented in Chapter 2, offset voltage V_{os} can usually be effectively nulled, leaving offset current I_{os} and the temperature drift of these quantities to be concerned with, dV_{os}/dT and dI_{os}/dT. In addition, the variation of these parameters with time, plus internal noise sources V_n referred to the amplifier input, must also receive serious consideration. Low-input-uncertainty requirements generally may be classified into those requiring low input bias current or low input voltage drift, with low-drift applications more prevalent. Low-input-bias-current amplifiers are required for current-output transducers such as piezoelectric devices, pH probes, and photomultipliers; typically they are JFET devices that have been optimized for I_{os}, although a varactor diode input amplifier such as the Teledyne Philbrick 1702 offers exceptional performance for commercially available devices with a bias current of 2 femtoamperes (2 × 10^{-15} A). Matched low-leakage varactors inherently eliminate $1/f$ noise, offer excellent offset and drift specifications, and provide 3 × 10^{11} ohms differential input impedance. Internal ac amplification and transformer coupling are responsible for these specific device characteristics. However, the relatively large capacitance between varactor inputs can be a cause of stability problems. A compensating capacitor C_f of a few picofarads, shown in Figure 3-4, improves stability at the expense of some bandwidth loss.

Differential chopper-input amplifiers are available for low-drift applications with an offset voltage drift of 0.3 $\mu V/°C$ and an exceptional 1 $\mu V/$ month, such as the monolithic Harris HA-2900, which employs frequent internal sampling and self-zeroing of error voltages with an output sample-hold amplifier for signal continuity during the error-correction interval. This device offers low-input-uncertainty characteristics similar to those of the previously discussed high-performance instrumentation amplifier, but it is

Figure 3-4 Inverting Varactor Charge Amplifier

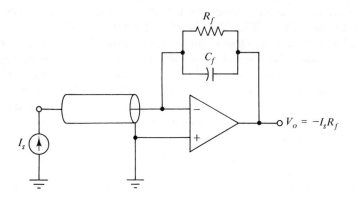

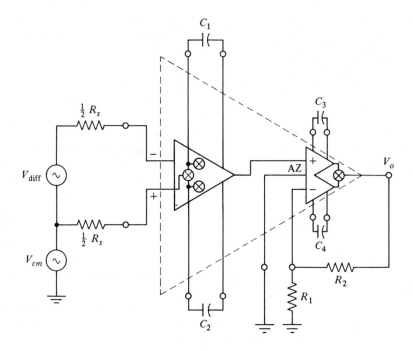

Figure 3-5 CAZ Instrumentation Amplifier

limited by the requirement for an external resistor network like the single-amplifier device of Figure 3-1. This differential chopper amplifier is well suited for precision integrators and bridge measurements, however, or as a buffer for precision signal sources.

The commutating auto-zero (CAZ) instrumentation amplifier is of similar operation and provides the same low drift without the requirement for an external resistor network in signal-acquisition applications, although its bandwidth is restricted between dc and 10 Hz. The CAZ amplifier represented by the Intersil ICL 7605 device of Figure 3-5, however, operates on principles different from those of conventional instrumentation amplifiers. It is self-compensating for internal error voltages whether they are derived from temperature or supply fluctuations or are variable over a long term. This device has one more input than a conventional operational amplifier, the auto-zero terminal, which is connected to the reference level of interest such as ground. A voltage converter commutates the differential input to a single-ended output, which is then supplied to the switched CAZ dual-operational amplifier. In a mode similar to the chopper-stabilized amplifier, these switched internal amplifiers and their associated capacitors alternately store and subtract the error voltages, referencing them to the noninverting input terminal. The differential voltage gain for this device is described by equation (3-9), and the selection of commutation capacitors is based on the manufacturer's recommendation. An internal oscillator and logic circuitry provide the commutation switching signals.

$$A_{v_{\text{diff}}} = \frac{R_1 + R_2}{R_1} \qquad (3\text{-}9)$$

When possibilities exist for troublesome ground disturbances or when signal-acquisition circuitry must be protected from fault conditions, an isolation amplifier is indicated. Such an amplifier permits a fully floating transducer loop and very high common-mode voltages to exist. A specialized instrumentation amplifier with conventional specifications, it includes an internal isolation barrier across which the signal and front-end power are coupled. This isolation barrier typically offers a resistance of 10^{11} ohms shunted by 10 pF. It is useful both for industrial interfacing with V_{iso} ratings to 5000 volts peak and to meet biomedical requirements for leakage currents of only a fraction of a microampere. In addition, most designs include a high-value series input resistor to limit input current in the event of input faults and amplifier failure.

The isolation voltage rating exists between the input and output commons shown in Figure 3-6, and the amplitude error referred to the output due to V_{iso} is determined by the amplifier isolation mode rejection ratio (IMRR) shown in equation (3-10). The $A_{v_{\text{diff}}}$ and CMRR of this equation pertain to the isolated differential input stage, and the output stage functionally is a unity-gain amplifier. Bias current required for operation of the input stage is internally supplied, so that no external dc restoration connections to ground or the power-supply common are required as with an instrumentation amplifier. Isolation amplifiers are especially useful in systems that must operate in very noisy environments including electromagnetic radiation.

LED-phototransistor-coupled and transformer-coupled circuits comprise presently available devices. Higher linearity and isolation performance generally accompany the transformer-coupled modulated-carrier method represented by the Analog Devices AD288 device, but bandwidth is limited to a few kilohertz. These specifications are sacrificed somewhat to economy in optical-coupled circuits represented by the Burr Brown 3650 amplifier. Typical IMRR ratings of 10^8 are available at dc and CMRR of 10^5. Isolated power is often provided to the amplifier front end by an internal dc/dc converter. An isolated power output may also be provided useful for a signal-conditioning preamplifier preceding the isolation amplifier.

$$V_o = A_{v_{\text{diff}}} \cdot V_{\text{diff}}(1 + \frac{1}{\text{CMRR}} \cdot \frac{V_{cm}}{V_{\text{diff}}}) + \frac{V_{\text{iso}}}{\text{IMRR}} \qquad (3\text{-}10)$$

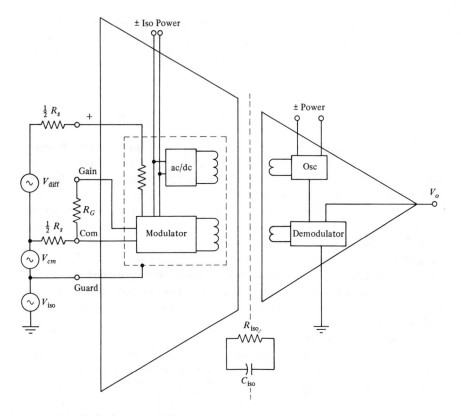

Figure 3-6 Isolation Amplifier

High-speed data-conversion and signal-conditioning circuits capable of accommodating pulse-type or video signals require wide-bandwidth operational amplifiers. These amplifiers are normally optimized for the most significant dynamic parameter of settling time because it embraces all of the other speed parameters of delay, rise time, slew rate, and transient subsidence to within a stated percent error. Figure 3-7 defines settling time as the maximum delay between the application of a step input and the instant that the output voltage swing remains within the stated tolerance range bracketing the equilibrium output voltage. Settling times to 0.1% of the full-scale amplifier output voltage in a fraction of a microsecond are typical, such as available with the Burr Brown 3400 device, with unity-gain bandwidths to 100 MHz and slew rates of 1000 V/μs. The output damping achieved in practice is principally a function of external R, L, and C component combinations. Parasitic reactive elements and/or carelessly planned circuit layouts can result in instability and an oscillatory tendency. Usually, a compensating trimmer capacitor C_f in the 5- to 20-pF range connected in

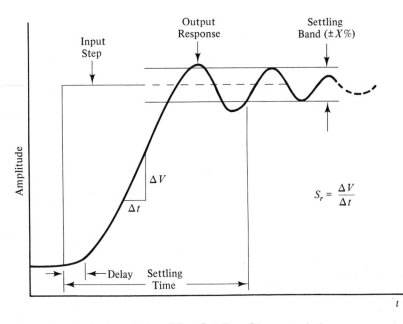

Figure 3-7 Wideband-Amplifier Settling Characteristics

accordance with Figure 3-8 can be adjusted to achieve an optimum response with minimum overshoot or ringing.

Amplifier slew rate depends directly upon the product of the output voltage amplitude and signal frequency, and this product must not exceed the slew-rate rating of the amplifier in any combination. Of some interest is the fact that the source of possible slew-induced distortion is the narrow range of input linearity (±25 mV) of the differential input stage. Aggravation by signal

Figure 3-8 Amplifier Output Loading Effects

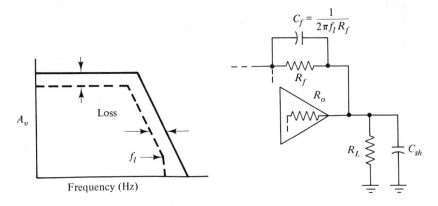

TABLE 3-2
EXAMPLE INSTRUMENTATION-AMPLIFIER SPECIFICATIONS

	Single Amplifier AD542	Three Amplifier AD522	High Performance LH0038	Low Bias Varactor TP1702	Low Drift Chopper HA2900	CAZ DC Amplifier ICL7605	Transformer Isolation AD288	Fast Settling BB3400
V_{os}	0.5 mV	0.2 mV	0.1 mV	5 mV	20 µV	2 µV	1 mV	1 mV
$\dfrac{dV_{os}}{dT}$	5 µV/°C	2 µV/°C	0.2 µV/°C	30 µV/°C	0.3 µV/°C	0.01 µV/°C	2 µV/°C	$\overline{50}$ µV/°C
I_{os}	2 pA	10 nA	7 nA	2 fA	50 pA	150 pA	50 pA	100 pA
$\dfrac{dI_{os}}{dT}$	X2/+10°C	50 pA/°C	0.5 nA/°C	1 fA/°C	1 pA/°C	1 pA/°C	0.1 nA/°C	X2/+10°C
S_r	3 V/µs	0.1 V/µs	0.3 V/µs	0.003 V/µs	2.5 V/µs	0.5 V/µs	0.1 µV/µs	1000 V/µs
$6.6 V_n \sqrt{10\,\text{Hz}}$ p-p	2 µV	1.5 µV	0.2 µV	10 µV	1 µV	4 µV	$\overline{1.5}$ µV	5 µV
A_{vo}	3×10^5	10^6	10^6	10^5	5×10^8	10^5	10^5	10^4
CMRR (IMRR)	10^4	10^5	10^6	10^5	10^6	10^5	$10^5 (10^6)$	10^3
f_{hi} @ $10^3 A_v$	700 Hz	200 Hz	10 kHz	1 Hz	3 kHz	10 Hz	200 Hz	100 kHz
$f(A_v)$	100 ppm	50 ppm	.1 ppm	1000 ppm	10 ppm	10 ppm	500 ppm	1000 ppm
$\dfrac{dA_v}{dT}$	R tempco	70 ppm/°C	7 ppm/°C	R tempco	R tempco	15 ppm/°C	35 ppm/°C	R tempco
$R_{i_{CM}}$	10^{11} Ω	10^9 Ω	10^9 Ω	10^{14} Ω	5×10^5 Ω	10^{12} Ω	10^8 Ω	10^{11} Ω

TABLE 3-3

AMPLIFIER DEFINITIONS

V_{os}	Input offset voltage
$\dfrac{dV_{os}}{dT}$	Input-offset-voltage temperature drift
I_{os}	Input offset current
$\dfrac{dV_{os}}{dT}$	Input-offset-current temperature drift
R_i	Input impedance
$R_{i_{diff}}$	Differential input impedance
$R_{i_{cm}}$	Common-mode input impedance
S_r	Slew rate
V_n	Input-referred noise voltage/$\sqrt{\text{Hz}}$
I_n	Input-referred noise current/$\sqrt{\text{Hz}}$
A_{v_o}	Open-loop gain
$A_{v_{cm}}$	Common-mode gain
$A_{v_{diff}}$	Closed-loop differential gain
$f(A_v)$	Gain nonlinearity
$\dfrac{dA_v}{dT}$	Gain temperature drift
f_{hi}	-3 dB bandwidth
CMRR (IMRR)	Common-mode (isolation-mode) intrinsic rejection ratio

integration provided by a capacitor-compensated high-gain innerstage results in the occurrence of this distortion on the rising and falling edges of waveforms. In all cases the amplifier open-loop bandwidth must exceed that of the input signal to preclude this distortion. Also, the identity of equation (3-11) must be satisfied. For example, a 1-V peak-to-peak sine-wave signal required at a signal frequency of 3 MHz specifies a minimum amplifier slew-rate requirement of 9.45 V/μs. If the amplifier is loaded by 1000 pF of output capacitance, then it must also be capable of delivering 10 mA of output current at this frequency. These relationships are illustrated by Figure 3-9. Interpolation of this figure shows that the Burr Brown 3400 is capable of driving 20 pF at 100 MHz, or 200 pF at 10 MHz, with a 20 mA output-current capability.

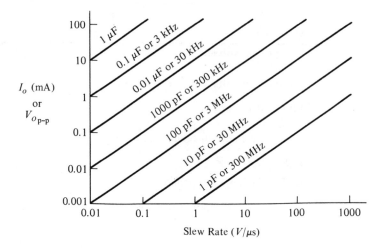

Figure 3-9 Amplifier Slew-Rate Curves

Wideband amplifiers are ordinarily operated in the inverting mode for best performance.

$$S_r = V_{o_{p-p}} \cdot \pi \cdot f_{\text{signal}} \qquad (3\text{-}11)$$

$$= \frac{I_o}{C_{sh}} \quad \text{V/s}$$

Amplifier output errors result primarily from output loading conditions and slew-rate limitations. Excessive output loading from a low output impedance R_L tends to reduce the available open-loop gain. From the Bode diagram of Figure 3-8 this results in a bandwidth loss; the problem can be overcome with a higher power output operational amplifier. When driving shielded cables or other capacitive loads, the parallel combination of the associated impedances forms an additional high-frequency corner f_l predictable from their time constant. Elimination of any resulting oscillatory tendency may be achieved with addition of the feedback capacitor C_f.

3-2 NOISE IN LOW-LEVEL AMPLIFICATION

An analysis of low-level signal amplification requires examination of all factors acting upon the input circuit and amplifier in order to account for the sources of nonvalid information introduced. Internal noise sources are important to the amplifier input uncertainty and involve both the transducer loop and amplifier. External interference and ground-loop problems are dealt with elsewhere—together with shielding practice in Chapter 6 and amplifier CMRR considerations in the previous section of this chapter. Two of the three significant internal noise sources are associated with the transducer loop: thermal noise and contact noise illustrated in Figure 3-10. The instrumentation amplifier provides the third internal noise source modeled as an equivalent noise voltage V_n and current I_n at the device input. Available circuit choices that affect these noise sources are consequently shown to be important in determining measurement accuracy in instrumentation applications through the internal noise levels provided.

The quantification of phenomena associated with the measurement of a quantity usually involves energy conversion, such as the thermal-to-electrical conversion of a thermocouple, plus energy-matter interactions, which produce additive noise. A measurement is therefore actually an estimate whose accuracy depends upon the minimization of this noise. Thermal noise V_t is present in all elements containing resistance; it is produced by the thermal agitation of electrons within the resistance. Equation (3-12) defines thermal-noise voltage as being proportional to the square root of both the source temperature and resistance. It is advantageous, therefore, to minimize both these quantities in order to minimize thermal noise per unit of root-Hertz bandwidth. Of course, this bandwidth must be large enough to provide for the necessary signal spectral occupancy described in Table 1-2 of Chapter 1.

Figure 3-10 Internal Transducer-Loop Noise Sources

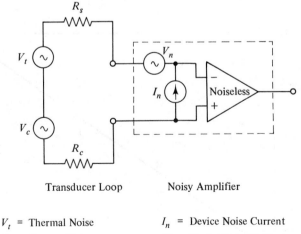

Transducer Loop   Noisy Amplifier

V_t = Thermal Noise I_n = Device Noise Current
V_c = Contact Noise V_n = Device Noise Voltage
R_c = Contact Resistance R_s = Source Plus Lead Resistance

Thermal noise is presented graphically by Figure 3-11 for room-temperature conditions (290°K), upon which moderate temperature variations have minor effect. Figure 3-10 represents this random noise source as a voltage generator in series with a noiseless source resistance R_s, the latter including both the transducer and connecting lead resistance.

$$V_t = \sqrt{4kTR_s} \quad \text{V}/\sqrt{\text{Hz}} \text{ rms} \qquad (3\text{-}12)$$

where $k = $ Boltzmann's constant (1.67×10^{-23} J/°K)
$T = $ absolute temperature (°K)
$R_s = $ source resistance (ohms)

A 1-k carbon composition resistor produces the same thermal noise as a 1-k wire-wound resistor at zero current flow. However, the carbon resistor with a dc current adds substantially more noise, owing to fluctuating conductivity from imperfect contact between the carbon particles. This contact noise occurs at any connection of conductors, including switches and semiconductor bonded contacts, and is usually referred to as $1/f$ or low-frequency-dependent noise. Owing to its frequency characteristic, contact noise is usually the dominant noise source at low frequencies in instrumentation systems. Contact noise, expressed as a noise voltage per root Hertz of bandwidth, is described by equation (3-13) and is of consequence primarily below a few

Figure 3-11 Thermal Noise Normalized to 1 Hz; $T = $ 290°K, *BW* = 1 Hz, $k = $ 1.67 × 10⁻²³ J/°K

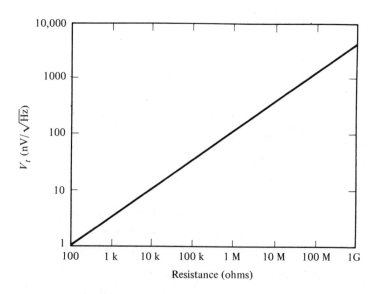

hundred Hertz. Above this frequency thermal noise is the primary circuit-internal-noise contribution. An important conclusion is that for low-level measurements, especially below 100 Hz, direct current flow through the transducer circuit should be either avoided or minimized. This requirement can be met by the use of low-input-bias-current instrumentation amplifiers such as those that employ FET preamplifiers. Equation (3-13) is a worst-case expression in that it assumes the noisiest contact and source resistance geometry and composition, such as that provided by carbon composition resistors:

$$V_c = (10^{-6}) \cdot \sqrt{\frac{1}{f}} \cdot (R_c + R_s)I_{dc} \quad V/\sqrt{Hz} \text{ rms} \qquad (3\text{-}13)$$

where I_{dc} = average dc current (A)
    f = signal frequency (Hz)
    R_s = source resistance (ohms)
    R_c = contact resistance (ohms)

The ultimate limitation to low-input-uncertainty amplification is attributable to thermoelectric effects resulting from electrical junctions of various metals and alloys. Unless all junctions are at precisely the same temperature, small thermoelectric voltages will be produced, typically of about 0.1 μV/$^\circ$C, although they can range to several tens of microvolts per degree centigrade for specific thermocouple materials. Thermoelectric solder (70% cadmium and 30% tin) should be used for low-level circuits. Precautions should also be taken to avoid amplifier thermal gradients by operating into a high-output-impedance load, and all devices should be enclosed to eliminate air movement across device surfaces. Carbon and metal film resistors produce larger thermocouple errors than wirewound resistors of Evenohm or Managinin, which generate 2 μV/$^\circ$C referenced to copper.

To specify noise for their devices, operational-amplifier manufacturers typically use the method of combined noise-voltage and noise-current sources applied to one input, as shown in Figure 3-10. The typical variation in these sources as a function of frequency is described by the curves of Figure 3-12. The equivalent short-circuit input rms noise voltage V_n is simply the random fluctuation that would appear to originate at the input of the noiseless amplifier if the input terminals were shorted. It is expressed in nV/$\sqrt{Hz}$ at a specified frequency, and its increase below 100 Hz is due to device internal $1/f$ contact noise sources. The equivalent open-circuit rms noise current I_n is the noise that apparently occurs at the noiseless amplifier input owing solely to noise currents. It is expressed in pA/$\sqrt{Hz}$ at a specified frequency, increases at lower frequencies with bipolar devices, and is essentially zero for FET devices. The combined transducer-circuit and amplifier-input noise sources add at the amplifier input in rms fashion, neglecting possible correlation in

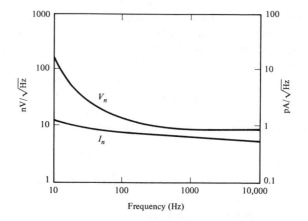

Figure 3-12 Amplifier RMS Noise Voltage and Current

their random fluctuations. The total rms noise voltage V_N is given by equation (3-14), which is convertible to p-p noise at 3.3σ confidence (0.1% error) with multiplication by 6.6 $\mu V_{p\text{-}p}/\mu V$ rms to account for its crest factor.

By way of example, the noise of a transducer circuit connected to a bipolar amplifier from Figure 3-12 is evaluated at the signal frequencies f of 10 Hz and 1 kHz, and R_s values of 100 ohms and 10 k. The amplifier has a typical 50-nA input bias current, and a negligible transducer contact resistance R_c is assumed. Equation (3-14) is represented on a root-Hz bandwidth basis in order to illustrate the effect of amplifier bandwidth, in this example 100 Hz and 10 kHz, on total internal noise V_N. These results are tabulated in Table 3-4, which shows amplifier noise voltage V_n to be dominant at low frequencies. Below about 100 Hz both bipolar and FET amplifier selection should

TABLE 3-4
TRANSDUCER LOOP PLUS AMPLIFIER INTERNAL NOISE

SIGNAL f (Hz)	AMPLIFIER f_{hi} (Hz)	TRANSDUCER R_s (ohms)	
10	100	100	
		10 k	
	10 k	100	
		10 k	
1 k	100	100	
		10 k	
	10 k	100	
		10 k	

therefore be based on a low V_n. At higher signal frequencies (1 kHz) total internal noise V_N is more sensitive to the source resistance R_s. Bipolar amplifiers exhibit a lower V_N at low values of R_s above 100 Hz than FET amplifiers, while FET types are a better choice at high R_s values because their I_n is essentially zero. Above source resistances of about 1 k and signal frequencies of 100 Hz, transducer circuit noise begins to dominate V_N. The magnitude of V_N is directly proportional to the amplifier bandwidth f_{hi} in all cases. Wide bandwidths and large source resistances result in the worst-case noise conditions at low signal frequencies; thus bandwidth should be limited where possible. For this purpose many instrumentation amplifiers include provision for bandwidth adjustment. The contact-noise contribution V_c is not significant in usual instrumentation applications because of the very small dc current flow involved.

$$V_N = (V_t^2 + V_c^2 + V_n^2 + I_n^2 \cdot R_s^2)^{1/2} \qquad (3\text{-}14)$$

$$\cdot \sqrt{f_{hi}} \quad \text{V rms} \quad (\times 6.6 \text{ for p-p})$$

Internal noise V_N is amplified by the voltage gain of the amplifier, so additional internal noise sources encountered in the instrumentation channel following the input stage generally are of negligible consequence. We can formally substantiate this observation by examining the noise contributed by the amplifier alone in terms of its noise figure F, defined by equation (3-15). This is expressed as the ratio of the total transducer-circuit noise power referred to the amplifier input, to the input noise power with no amplifier present. Note that contact noise V_c cannot contribute to the transducer-circuit noise with the amplifier removed because of the absence of dc current flow. The overall noise figure F_T for a network of n amplifiers in cascade was shown by Friis (Reference 17) to be essentially determined by the noise figure F_1 and

TRANSDUCER ($nV/\sqrt{Hz}$)		AMPLIFIER ($nV/\sqrt{Hz}$)		Total (μV)
V_t	V_c	V_n	$I_n \cdot R_s$	$V_N\,p\text{-}p$
1.2	0.0016	150	0.15	9.9
12.0	0.16	150	15.0	10.6
1.2	0.0016	150	0.15	99.0
12.0	0.16	150	15.0	106.0
1.2	0.00016	9	0.07	0.6
12.0	0.016	9	7.0	1.0
1.2	0.00016	9	0.07	59.0
12.0	0.016	9	7.0	109.0

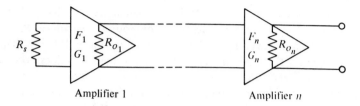

Figure 3-13 Cascaded Amplifier Network

available power gain G_1 of the first stage. Consequently, the noise performance of the first stage is important because it determines the noise of the instrumentation channel. Available power gain is the square of the voltage gain multiplied by the ratio of the source resistance to the internal output impedance for each amplifier $[G = A_v^2 \cdot R_s/R_o]$. A noiseless amplifier has a noise figure of $F = 1$, and typical bipolar and FET operational amplifiers are in the range of $F = 2$ to 3 depending upon source resistance and signal frequency. This relationship is illustrated by Figure 3-13 and equation (3-16).

The effects of the input noise voltage and current on an instrumentation amplifier combine to limit the signal sensitivity of the amplifier. Total internal noise V_N is clearly minimized for minimum R_s values. However, in terms of equations (3-12) and (3-15) amplifier noise figure F approaches infinity as R_s approaches zero. This contradiction results from an overgeneralization of noise performance indicated by noise figure, so that a reliable comparison is not provided between various circuit conditions for the same amplifier. Therefore, for determining noise performance of instrumentation amplifiers, reliance on V_n and I_n is more useful.

$$F = \frac{V_t^2 + V_c^2 + V_n^2 + I_n^2 \cdot R_s^2}{V_t^2} \qquad (3\text{-}15)$$

$$F_T = F_1 + \frac{F_2 - 1}{G_1} + \cdots + \frac{F_n - 1}{G_1 \cdots G_{n-1}} \qquad (3\text{-}16)$$

3-3 AMPLIFIER ERROR ANALYSIS

Instrumentation amplifiers are generally the first active device encountered at the input of analog signal-acquisition systems. They are typically used to interface electrical transducers to computer analog input systems, and to electronic controllers or recorders in process-control and data-logging applications. It is obvious from the instrumentation-amplifier parameters presented in Table 3-2 that these components offer individual performance characteristics that are optimized for specific applications. Nevertheless, it is essential to perform an error analysis in order to identify the significant independent sources and their contribution in specific applications, and to apply a root-sum-squared summation to these sources to determine the amplifier error budget referred to its input. Such a procedure is developed from equation (1-4), and the resulting composite error is described by equations (3-18) and (3-19) related to the V_{FS} output value.

The choice of an instrumentation amplifier is based upon a selection of parameters that will minimize the error budget for that application. Criteria often include the amplifier's ability to reject common-mode interference, for example encountered with a bridge-circuit transducer or from induced noise, and the minimization of amplifier input uncertainty due to its offset, noise, and drift parameters. The $V_{error_{RTI}}$ composite input-referred error may then be multiplied by $A_{v_{diff}}$ to determine amplifier error as a percentage of the full-scale output amplitude. The significance of this is that the amplifier error substantially influences the signal-measurement accuracy of the total system to which the amplifier is connected. Consequently, single-amplifier instrumentation amplifiers are usually employed in noncritical applications because of the difficulty associated with their external resistor networks. For similar reasons the three-amplifier instrumentation amplifier is preferably chosen as an integral, manufacturer-trimmed component. However, for comparison purposes we will design a single-amplifier instrumentation amplifier using the Analog Devices AD542 device described in Table 3-2.

The majority of instrumentation-amplifier applications are at low frequencies because of the predominant low-frequency response of the physical processes from which measurements are typically sought. Transducer elements frequently are either bridge circuits and strain gages, which require a high CMRR for rejection of the common-mode excitation voltage, or low-level signal sources such as thermocouples, which require a very low amplifier input offset and drift uncertainty to preserve measurement accuracy. And to the extent possible the realization of both high CMRR and low input uncertainty is ideal for most applications, except perhaps those requiring very low amplifier bias current or fast-settling response. In addition, induced common-mode interference is dominated by 60-Hz sources in both laboratory and industrial environments because of the power load at this line frequency.

In applications that require very high common-mode rejection, parasitic capacitances at the amplifier inputs can significantly degrade CMRR if they are unbalanced. This is especially a problem with long shielded transducer cables because of unequal line-to-shield capacitance values. The equivalent circuit can be viewed as a bridge, shown in Figure 3-14, that is unbalanced by the differences in source impedances and line-to-shield capacitances. The maximum realizable CMRR due to this external amplifier circuit imbalance is defined by equation (3-17), and is normally evaluated at 60 Hz, since this is

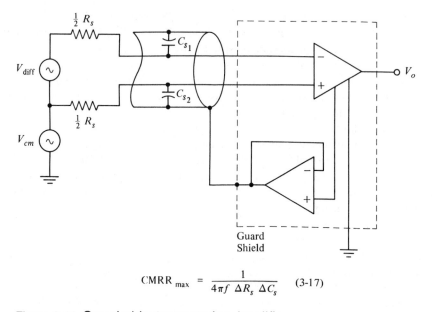

$$\text{CMRR}_{\max} = \frac{1}{4\pi f \; \Delta R_s \; \Delta C_s} \quad (3\text{-}17)$$

Figure 3-14 Guarded Instrumentation Amplifier

the dominant interference frequency. In practice, R_s and C_s can be measured at the amplifier input to determine the maximum available CMRR. For example, a 25-ohm resistance imbalance and 50-pF input capacitance imbalance result in a 10^6 maximum realizable CMRR.

Input guarding resolves the capacitance-imbalance problem by driving the cable shield with the derived common-mode signal, which forces the shield and inner conductors to track the common-mode interference. This signal is frequently made available with high-performance instrumentation amplifiers at a guard-drive terminal. The guard-drive buffer of Figure 3-14 lowers the drive-source impedance to further boost CMRR. However, in practice realizable values of CMRR even with guarded input circuits are limited to about 10^6. It should also be noted that CMRR rolls off with the open-loop gain characteristic of an instrumentation amplifier, and the maximum value is usually preserved only between dc and 100 Hz.

It is useful to examine the specifications of instrumentation amplifiers representing a range of performance under identical operating conditions. For a differential voltage gain $A_{V_{\text{diff}}}$ of 10^3 and a nominal temperature variation dT of $20°C$, combined amplifier error budgets are derived for five instrumentation amplifiers and tabulated in Table 3-5 for a low-level V_{diff} of 10 mV dc and 1 V rms V_{cm}. The potentiometric feedback single-amplifier version achieves high gain without abnormally large feedback resistors, employing equations (2-14), (3-3), and (3-18), and Table 3-1 to determine its parameters. This circuit allows an R_i of 500 k for insignificant input-signal loading.

However, to preserve the in-circuit CMRR the source resistance must be in balance to within 50 ohms with this R_i value. Gain drift dA_v/dT for this example is essentially due to the external resistor temperature-coefficient tracking and as considered here is 100 ppm/°C. All of the amplifier error budgets consist of both systematic and random errors such as offset and noise, respectively. Although additional minor terms may be present, only the error terms that are significant to the error budget of present-day devices have been included. The in-circuit CMRR values are calculated for the single- and three-amplifier examples shown in Figures 3-15 and 3-16. CMRR values for the remaining instrumentation amplifiers are taken as the 10^5 and 10^6 intrinsic values from Table 3-2.

Figure 3-15 Potentiometric Feedback Subtractor Instrumentation Amplifier

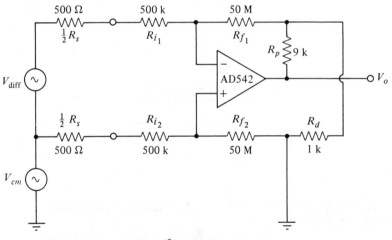

(1% tolerance, 100 ppm/°C tempco resistors)

$$A_{v_{\text{diff}}} = \frac{R_f + R_p + R_f \cdot R_p/R_d}{R_i} \qquad (3\text{--}20)$$

$$= \frac{50\,\text{M} + 9\,\text{k} + 50\,\text{M} \cdot 9\,\text{k}/1\,\text{k}}{500\,\text{k}} = 10^3$$

$$\text{CMRR} = \frac{A_{v_{\text{diff}}}}{A_{v_{cm}}} = \frac{10^3}{0.02} = 5 \times 10^4 \qquad (2\text{--}14)$$

$$\text{CMRR}_{\text{in-circuit}} = \frac{\text{CMRR} \cdot \text{CMRR}_{\text{intrinsic}}}{\text{CMRR} + \text{CMRR}_{\text{intrinsic}}} \qquad (3\text{--}3)$$

$$= \frac{5 \times 10^4 \cdot 10^4}{5 \times 10^4 + 10^4} = 0.833 \times 10^4$$

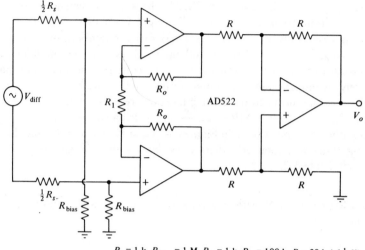

$$R_s = 1 \text{ k}, R_{bias} = 1 \text{ M}, R_1 = 1 \text{ k}, R_o = 100 \text{ k}, R = 20 \text{ k} \ (\leq \tfrac{1}{2} \%)$$

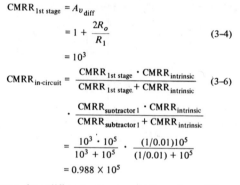

$$\text{CMRR}_{\text{1st stage}} = A v_{\text{diff}}$$

$$= 1 + \frac{2R_o}{R_1} \qquad (3\text{-}4)$$

$$= 10^3$$

$$\text{CMRR}_{\text{in-circuit}} = \frac{\text{CMRR}_{\text{1st stage}} \cdot \text{CMRR}_{\text{intrinsic}}}{\text{CMRR}_{\text{1st stage}} + \text{CMRR}_{\text{intrinsic}}} \qquad (3\text{-}6)$$

$$\cdot \ \frac{\text{CMRR}_{\text{subtractor1}} \cdot \text{CMRR}_{\text{intrinsic}}}{\text{CMRR}_{\text{subtractor1}} + \text{CMRR}_{\text{intrinsic}}}$$

$$= \frac{10^3 \cdot 10^5}{10^3 + 10^5} \cdot \frac{(1/0.01)10^5}{(1/0.01) + 10^5}$$

$$= 0.988 \times 10^5$$

Figure 3-16 Three-Amplifier Instrumentation Amplifier with Floating Transducer

An important consideration of amplifier gain is the nonlinearity of the gain function $f(A_v)$ rather than the accuracy of the gain formula. Any inaccuracy in the gain formula merely results in a scaling error, which can be compensated by the gain adjustment. However, gain nonlinearity is not easily compensated, and it results in an amplitude error over the dynamic range of the signal. This is compounded by the gain temperature drift dA_v/dT produced by active and passive amplifier components, which also contributes to the amplifier input uncertainty. Note that floating transducer-amplifier inputs require dc-restoration resistors R_{bias}, as shown in Figure 3-16. These values are normally chosen equal or greater than $10^3 \ R_s$ and 0.1% tolerance to minimize both input voltage-divider effects and CMRR derogation.

The results of the calculations shown in Table 3-5 illustrate that some choice exists between optimizing on CMRR or the dc drift parameters with the various instrumentation amplifiers. Offset-voltage temperature drift is a major contributor to the error budgets, offering the designer an incentive to provide a benign thermal environment. Offset voltage is considered nulled in these examples. At the low signal frequency and source resistance involved the total transducer-loop noise V_n p-p is dominated by the amplifier noise component V_n for the AD522, AD288, and ICL7605 amplifiers, and is taken from the noise entries in Table 3-2. The 1 M total R_i and 700-Hz bandwidth of the AD542 circuit result in a dominant transducer-loop noise V_t of 21 μV p-p from equation (3-12), which is similar to that of the 10-kHz-bandwidth LH0038 amplifier and its R_s of 1 k that also produce a dominant transducer-loop noise V_t of 3.3 μV p-p.

Although the single-amplifier version is capable of useful performance, implementation of the external resistors and their compromises limit application of this circuit to less critical applications. The performance and value of

TABLE 3-5

INSTRUMENTATION-AMPLIFIER ERROR BUDGETS

$V_{\text{diff}} = 10$ mV dc, $V_{cm} = 1$ V rms, $dT = 20°C, A_{v\,\text{diff}} = 10^3, R_s = 1$ k, $V_{\text{FS output}} = 10$ V)

Parameter	Single Amplifier AD542	Three Amplifier AD522	Isolation Amplifier AD288	CAZ Amplifier ICL7605	High Performance LH0038
$\dfrac{dV_{os}}{dT} \cdot dT$	100 μV	40 μV	40 μV	0.2 μV	4 μV
$I_{os} \cdot R_s$	0.002	10	0.05	0.15	7
$\dfrac{dI_{os}}{dT} \cdot R_s \cdot dT$	0.008	1	2	0.02	10
V_N p–p	21	1.5	1.5	4	3.3
$\dfrac{\sqrt{2} \cdot V_{cm}}{\text{CMRR}_{\text{in-circuit}}}$	170	14	14	14	1.4
$f(A_v) \cdot \dfrac{V_{\text{FS output}}}{A_{v\,\text{diff}}}$	1	0.5	5	0.1	0.01
$\dfrac{dA_v}{dT} \cdot dT \cdot \dfrac{V_{\text{FS output}}}{A_{v\,\text{diff}}}$	20	14	7	3	1.4
RMS $V_{\text{error}\,\text{RTI}}$	198 μV	46 μV	43 μV	15 μV	13 μV
$\epsilon_{\text{ampl}\,\%\text{FS}}$	1.98%	0.46%	0.43%	0.15%	0.13%

the three-amplifier version containing an internal resistor network trimmed by the manufacturer is especially apparent from this error analysis. And although bipolar amplifiers generally offer lower V_{os} and dV_{os}/dT than FET devices, generalizations about when to apply which type are of minimal utility in comparison with an evaluation of their error budgets. Therefore, such analyses are essential for comparing amplifier performance for specific applications under prevailing conditions, and for identifying where error reduction is most productive in minimizing their combined errors. Peak V_{cm} attenuated by amplifier CMRR is referred to the amplifier input by the computation shown in Table 3-5. The effects of both gain nonlinearity $f(A_v)$ and drift dA_v/dT are also referenced to the amplifier input. The rms combination of these errors provided by equation (3-18) is then normalized to the V_{FS} output level by equation (3-19). Complete instrumentation-amplifier application examples are presented in Sections 6-3 and 6-4 of Chapter 6. Note that $V_{error_{RTI}}$ of equation (3-18) determines the minimum input signal that can be confidently resolved, the threshold being where V_{diff} equals $V_{error_{RTI}}$. The $\sqrt{2}$ times V_{cm} is appropriate for typical sinusoidal rms interference.

$$V_{error_{RTI}} = \left\{ \left(\frac{dV_{os}}{dT} \cdot dT \right)^2 + (I_{os} \cdot R_s)^2 + \left(\frac{dI_{os}}{dT} \cdot R_s \cdot dT \right)^2 \right.$$

$$+ (V_{N_{p-p}})^2 + \left(\frac{\sqrt{2} \cdot V_{cm}}{CMRR} \right)^2 + \left(f(A_v) \cdot \frac{V_{FS\ output}}{A_{v_{diff}}} \right)^2$$

$$\left. + \left(\frac{dA_v}{dT} \cdot dT \cdot \frac{V_{FS\ output}}{A_{v_{diff}}} \right)^2 \right\}^{1/2} \qquad (3\text{-}18)$$

$$\epsilon_{ampl\%FS} = \frac{V_{error_{RTI}} \cdot A_{v_{diff}}}{V_{FS\ output}} \cdot 100\% \qquad (3\text{-}19)$$

REFERENCES

1. D. C. BAILEY, "An Instrumentation Amplifier Is Not an Op Amp," *Electronic Products*, September 18, 1972.

2. A. P. BROKAW, "Use a Single Op Amp for Many Instrumentation Problems," *Electronic Design News*, April 1, 1972.

3. M. CALLAHAN, "Chopper-Stabilized IC Op Amps Achieve Precision, Speed, Economy," *Electronics*, August 16, 1973.

4. J. W. JAQUAY, "Designers Guide to Instrumentation Amplifiers," *Electronic Design News*, May 5, 1973.

5. D. JONES and R. W. WEBB, "Chopper-Stabilized Op Amp Combines MOS and Bipolar Elements," *Electronics*, September 27, 1973.

6. J. H. KOLLATAJ, "Reject Common-Mode Noise," *Electronic Design*, April 26, 1973.

7. T. C. LYERLY, "Instrumentation Amplifier Conditions Computer Inputs," *Electronics*, November 6, 1972.

8. F. POULIST, "Simplify Amplifier Selection," *Electronic Design*, August 2, 1973.

9. J. I. SMITH, *Modern Operational Amplifier Circuit Design*, John Wiley, New York, 1971.

10. G. TOBEY, J. GRAEME, and L. HUELSMAN, *Operational Amplifiers: Design and Applications*, McGraw-Hill, New York, 1971.

11. C. F. WOJSLAW, "Use Op Amps with Greater Confidence," *Electronic Design*, March 16, 1972.

12. R. L. YOUNG, "Lift IC Op Amp Performance," *Electronic Design*, February 15, 1973.

13. J. G. GRAEME, *Applications of Operational Amplifiers: Third-Generation Techniques*, McGraw-Hill, New York, 1973.

14. J. A. CONNELLY, *Analog Integrated Circuits*, Wiley-Interscience, New York, 1975.

15. *Linear Applications Handbook 2*, National Semiconductor Corporation, 2900 Semiconductor Drive, Santa Clara, Calif.

16. S. DAVIS, "IC Op Amp Families," *Electronic Design News*, January 20, 1978.

17. H. T. FRIIS, "Noise Figures in Radio Receivers," *Proceedings IRE*, vol. 32, July 1944.

PROBLEMS

3-1. A single-amplifier instrumentation amplifier with only a source-impedance imbalance will result in a CMRR of twice that predicted by Table 3-1 on a percentage imbalance basis. Find the CMRR for an $A_{v_{\text{diff}}} = 100$ and a ΔR_s of 1 k with 50-k input resistors. Ignore resistor tolerance for this example.

3-2. A CRT display is driven from an amplifier over a 72-ohm coaxial cable. Determine the amplifier slew rate and output current requirements to provide 1 V p-p for 3-MHz video bandwidth.

3-3. A 10-mV dc signal in 1 V rms of common-mode noise is to be raised to a 1-V dc output level with 10 mV rms noise. Determine the required $A_{v_{\text{diff}}}$ and CMRR.

3-4. A 3.5-kHz-bandwidth bipolar instrumentation amplifier terminates a 1-k source-resistance bridge transducer circuit providing a 1-V dc common-mode voltage. Determine the total internal transducer-loop noise $V_{N_{p-p}}$ at a frequency of 10 Hz.

3-5. Determine the maximum realizable CMRR at 60 Hz for a transducer loop that exhibits a 950-ohm differential resistance and 14-pF differential capacitance imbalance.

3-6. A Teledyne Philbrick 1702 varactor amplifier is used to mechanize an electrometer that is maintained at its operating temperature within 1°C. Determine the output $\epsilon_{\text{ampl}\%\text{FS}}$ under the following operating conditions using the method of Table 3-5: $I_{\text{diff}} = 1\ \mu\text{A}$, $R_s = 1\ \text{k}$, $V_{cm} = 1\ \text{V rms}$, $dT = 1°\text{C}$, $A_{v_{\text{diff}}} = 10^3$, $V_{\text{FS}} = 1\ \text{V}$. Repeat for $I_{\text{diff}} = 10\ \mu\text{A}$ and $A_{v_{\text{diff}}} = 10^2$. All resistors are 100 ppm/°C tempco.

active filter
design

4-0 INTRODUCTION

A frequency-selective filter is an electrical network designed to pass signals in a specific frequency range coinciding with a given signal spectral occupancy, called the *passband*, and to attenuate signals at all other frequencies, called the *stopband*. Realizable filters are also characterized by a transition band between the passband and stopband, whose exact boundary locations depend upon the specific filter approximation. Amplitude $A_{(f)}$ and phase $\beta_{(f)}$ functions usually are employed to classify the various filter types, where the ideal filter has zero amplitude loss in the passband and infinite loss in the stopband with linear phase throughout. However, the ideal filter is physically unrealizable because practical filters are represented by ratios of polynomials that cannot possess the discontinuities required for sharply defined boundaries and a zero-width transition band. Realizable filters are therefore an approximation, to within some specified set of tolerances, to the ideal.

For instrumentation purposes filter requirements are usually lowpass in nature because of the typically encountered low-frequency signals and the general requirement to eliminate higher-frequency interference, plus the prevention of signal hetrodyning (aliasing) with the sampling frequency in sampled-data systems. However, highpass and bandreject filters are also occasionally required as well as the bandpass filter. The excellent behavior and competitive cost of active filters in the dc to 100-kHz instrumentation signal-frequency range make these mechanizations especially attractive for applications such as high-performance transducer-amplifier-filter signal conditioning. Consequently, this chapter develops accepted instrumentation-filter approximations, such as the Butterworth, Chebyshev, and Bessel lowpass and highpass responses, plus the characterization of bandpass and bandreject responses. Stable active filter networks are then introduced and practical realization methods presented, including design tabulations and illustrative implementation examples. A final section on filter-error analysis examines the sources of error and their quantitative contribution to an overall system, and specifies appropriate filter types for various signals and their passband spectral occupancy which minimize filter component error.

4-1 INSTRUMENTATION FILTER TYPES

In approximating the ideal lowpass filter response suggested by Figure 4-1 with realizable filter transfer functions, it is advantageous to structure them to correspond to known and well-behaved mathematical functions such as Butterworth and Chebyshev polynomials. The Butterworth amplitude-response function, described by equations (4-1) and (4-2) and Table 4-1, in all cases is seen to be specified by its order n, or the number of poles of the filter. Figure 4-2 describes the Butterworth lowpass amplitude response $A(f)$ and Figure 4-3 its phase response $\beta(f)$. Butterworth lowpass filters are characterized by a maximally flat amplitude response in the vicinity of dc, which extends toward its -3-dB cutoff frequency f_c as n increases. Their attenuation increases rapidly beyond f_c for large n, and their phase response is only slightly nonlinear, which provides a good approximation to the ideal case.

Figure 4-1 Ideal Lowpass Filter

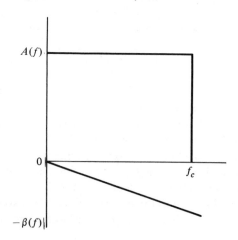

TABLE 4-1
BUTTERWORTH POLYNOMIAL COEFFICIENTS

POLES n	b_0	b_1	b_2	b_3	b_4	b_5
1	1.0					
2	1.0	1.414				
3	1.0	2.0	2.0			
4	1.0	2.613	3.414	2.613		
5	1.0	3.236	5.236	5.236	3.236	
6	1.0	3.864	7.464	9.141	7.464	3.864

Figure 4-4 presents the Butterworth highpass amplitude response, where Figures 4-2 through 4-5 were adapted from curves copyrighted by the Burr Brown Research Corporation.

$$A(f) = \frac{b_o}{\sqrt{B(s)B(-s)}} \tag{4-1}$$

$$= \frac{1}{\sqrt{1 + (f/f_c)^{2n}}}$$

Figure 4-2 Butterworth Lowpass Amplitude

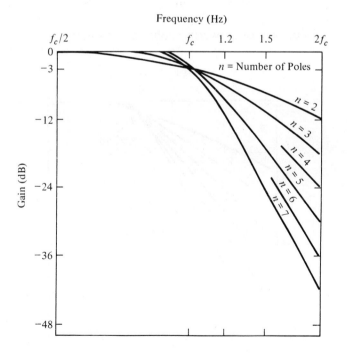

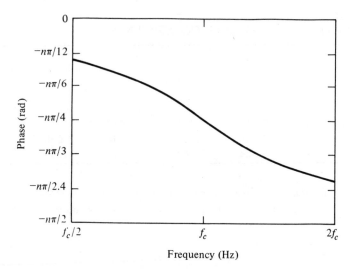

Figure 4-3 Butterworth Lowpass Phase

$$B(s) = \left(j\,\frac{f}{f_c}\right)^n + b_{n-1}\left(j\,\frac{f}{f_c}\right)^{n-1} + \cdots + b_0 \qquad (4\text{-}2)$$

Figure 4-4 Butterworth Highpass Amplitude

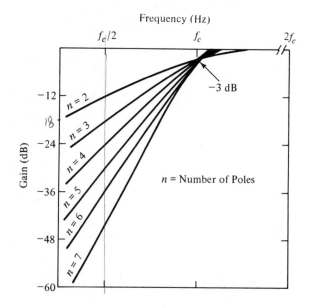

A disadvantage of the Butterworth response is that the zeros of its transfer function are all lumped at zero frequency, which results in a departure from the ideal response at f_c. In contrast, the zeros of the Chebyshev lowpass filter transfer function are spread out across the passband, forcing its amplitude response to attain a maximum value at a number of passband frequencies. As a result of this distribution, this filter is sometimes called an equiripple filter. The Chebyshev amplitude function is described by equation (4-3), and the ripple height by equation (4-4). For example, a 0.1-dB ripple passband corresponds to an $\epsilon = 0.15$. Note that an odd number of Chebyshev poles n results in the ripple's being at the maximum passband amplitude value at dc, and an even n places the ripple at this maximum value minus the ripple height at dc. Both odd and even n values pass through this latter amplitude value at the cutoff frequency f_c. Table 4-2 provides a tabulation for C_n through the sixth-order polynomial.

$$A(f) = \frac{1}{\sqrt{1 + \epsilon^2 C_n^2(f/f_c)}} \tag{4-3}$$

$$\text{ripple height} = 10 \log (1 + \epsilon^2) \tag{4-4}$$

Increasing ϵ for a fixed n increases the transition-band rate of attenuation, and hence the stopband performance, but creates a larger passband ripple. Increasing n for a fixed ϵ improves stopband performance without a larger passband ripple, but at the expense of additional filter complexity. For a given filter order n the Chebyshev lowpass provides a closer approximation to the ideal filter than the Butterworth lowpass at cutoff and in the stopband, albeit with some sacrifice in passband flatness. Chebyshev amplitude and phase responses are plotted for 1-dB passband ripple in Figures 4-5 and 4-6. Figure 4-7 provides a comparison of fifth-order Butterworth and Chebyshev lowpass filters for a Chebyshev passband ripple of 0.1 dB, which corresponds to 0.8% amplitude error.

TABLE 4-2
CHEBYSHEV POLYNOMIALS OF THE FIRST KIND

POLES n	$C_n(f/f_c)$
1	(f/f_c)
2	$2(f/f_c)^2 - 1$
3	$4(f/f_c)^3 - 3(f/f_c)$
4	$8(f/f_c)^4 - 8(f/f_c)^2 + 1$
5	$16(f/f_c)^5 - 20(f/f_c)^3 + 5(f/f_c)$
6	$32(f/f_c)^6 - 48(f/f_c)^4 + 18(f/f_c)^2 - 1$

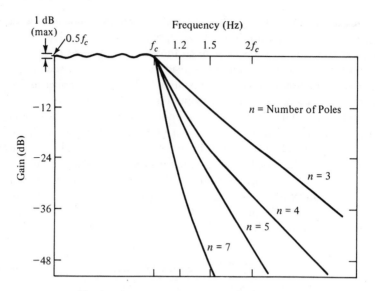

Figure 4-5 Chebyshev Lowpass Amplitude (1-dB ripple)

Figure 4-6 Chebyshev Lowpass Phase (1-dB ripple)

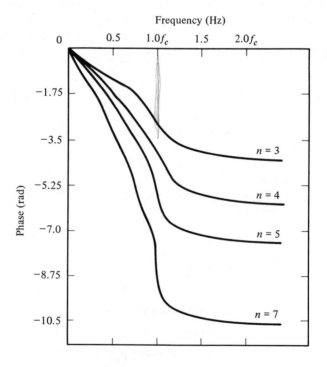

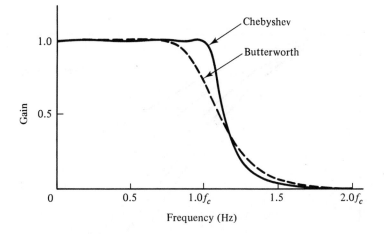

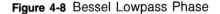

Figure 4-7 Five-Pole Lowpass Butterworth and Chebyshev (0.1-dB) Comparison

Bessel lowpass filters are characterized by a linear phase delay beginning at dc and extending to their cutoff frequency f_c and beyond as a function of filter order n shown in Figure 4-8. Unlike Butterworth filters, the attenuation at f_c varies with the number of poles n and has a Gaussian amplitude response described by Figure 4-9. Output amplitude overshoot to a

Figure 4-8 Bessel Lowpass Phase

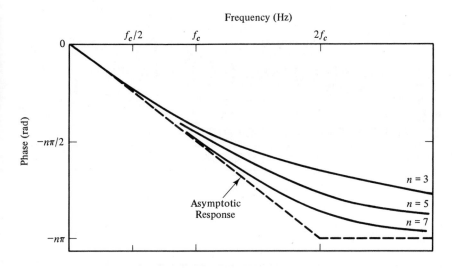

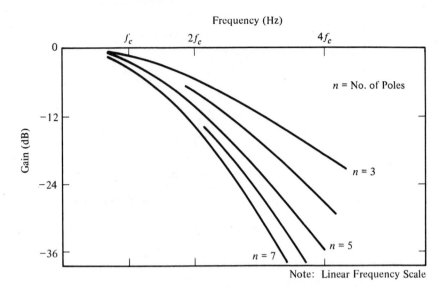

Figure 4-9 Bessel Lowpass Amplitude

step input is essentially zero for Bessel filters, where Chebyshev filters may exhibit more than 25% overshoot. Because of these characteristics Bessel filters can be employed for time delays, for filtering of pulse-type instrumentation signals such as from voltage-to-frequency converters, and for data-smoothing "running average" filters. The Bessel lowpass amplitude function is described by equations (4-1) and (4-2) up to a sixth-order approximation with the coefficients of Table 4-3.

$$A(f) = \frac{b_0}{\sqrt{B(s)B(-s)}} \qquad (4\text{-}1)$$

$$B(s) = \left(j\, \frac{f}{f_c} \right)^n + b_{n-1} \left(j\, \frac{f}{f_c} \right)^{n-1} + \cdots + b_0 \qquad (4\text{-}2)$$

TABLE 4-3
BESSEL POLYNOMIAL COEFFICIENTS

POLES n	b_0	b_1	b_2	b_3	b_4	b_5
1	1					
2	3	3				
3	15	15	6			
4	105	105	45	10		
5	945	945	420	105	15	
6	10,395	10,395	4725	1260	210	21

The bandpass filter passes a band of frequencies of bandwidth Δf centered at a frequency f_0 and attenuates all other frequencies. The quality factor Q of this filter is a measure of its selectivity and is defined by the ratio $f_0/\Delta f$. Also of interest is the geometric mean of the upper and lower -3-dB frequencies defining Δf, or $f_g = f_u \cdot f_\ell$. Equations (4-5) and (4-6) present the amplitude function for a second-order bandpass filter in terms of these quantities, with amplitude response for various Q values plotted in Figure 4-10.

It may be appreciated from this figure that for all Q values the bandpass skirt attenuation rolloff relaxes to -12 dB/octave one octave above and below f_0, which is expected for any second-order filter. (An octave is the interval between two frequencies, one twice the other.) Greater skirt attenuation can be obtained by cascading these single-tuned sections, thereby producing a higher-order filter. The phase response of a bandpass filter may be envisioned as that of a highpass and lowpass filter in cascade. This phase has a slope whose rate of descent is maximum and of value 0 degrees at f_0, asymptotically reaching its maximum positive and maximum negative phase shift below and above f_0, respectively; total phase shift is a function of the filter order n.

Figure 4-10 Bandpass Amplitude Response

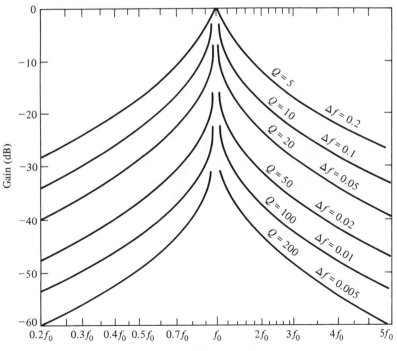

$$A(f) = \frac{2\pi f_0/Q}{\sqrt{B(s)B(-s)}} \qquad (4\text{-}5)$$

$$B(s) = (j2\pi f) + \frac{2\pi f_0}{Q} + \frac{(2\pi f_0)^2}{j2\pi f} \qquad (4\text{-}6)$$

The bandreject filter, also called a band-elimination or notch filter, passes all frequencies except those centered about f_c. Its amplitude function is described by equations (4-1), (4-7), and (4-8), and its amplitude response by Figure 4-11. Bandreject Q is determined by the ratio $f_c/\Delta f$, where bandwidth Δf is defined between the -3-dB passband cutoff frequencies. Bandreject-filter phase response follows the same phase characteristics described for the bandpass filter. For instrumentation service the bandreject response can be obtained from the lowpass Butterworth coefficients of Table 4-1, and a maximally flat passband can be realized with a lowpass-to-bandreject transformation.

$$A(f) = \frac{1}{\sqrt{B(s)B(-s)}} \qquad (4\text{-}1)$$

$$B(s) = C^n + b_{n-1}C^{n-1} + \cdots + b_0 \qquad (4\text{-}7)$$

$$C = \frac{\Delta f(j2\pi f)}{(j2\pi f)^2 + (2\pi f_c)^2} \qquad (4\text{-}8)$$

Figure 4-11 Bandreject Amplitude Response

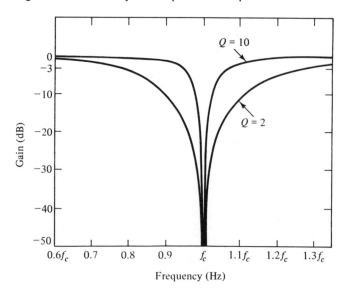

4-2 ACTIVE FILTER NETWORKS

In 1955 Sallen and Key[1] of MIT published a description of 18 active filter networks for the realization of various filter approximations. However, a rigorous sensitivity analysis by Geffe[2] and others disclosed by 1967 that only four of the original networks exhibited low sensitivity to component drift. Of these, the unity-gain and multiple-feedback networks are of particular value for implementing lowpass and bandpass filters, respectively, to Q values of 10. Work by others resulted in the low-sensitivity biquad resonator, which can provide stable Q values to 200, and the stable gyrator bandreject filter. These four networks are shown in Figure 4-12 with key sensitivity parameters. The sensitivity of a network can be determined, for example, when the change in its Q for a change in its passive-element values is evaluated. Equation (4-9) describes the change in the Q of a network by multiplying the thermal coefficient of the component of interest by its sensitivity coefficient. Normally, 50-to-100-ppm/°C components yield good performance.

$$S_C^Q = \pm \frac{1}{2} \qquad\qquad (4\text{-}9)$$

$$= (\pm \frac{1}{2})(50 \text{ ppm}/°C) \, (100\%)$$

$$= \pm 0.0025\% \, Q/°C$$

Unity-gain networks offer excellent performance for lowpass and highpass realizations and may be cascaded for higher-order filters. This is perhaps the most widely applied active filter circuit. Note that its sensitivity coefficients are less than unity for its passive components—the sensitivity of conventional passive networks—and that its resistor temperature coefficients are zero. However, it is sensitive to filter gain, indicating that designs that also obtain greater than unity gain with this filter network are suboptimum. The advantage of the multiple-feedback network is that a bandpass filter can be formed with a single operational amplifier, although the biquad network must be used for high Q bandpass filters. However, the stability of the biquad at higher Q values depends upon the availability of adequate amplifier loop gain at the filter center frequency. Both bandpass networks can be stagger-tuned for a maximally flat passband response when required. The principle of operation of the gyrator is that a conductance $-G$ gyrates a capacitive current to an effective inductive current. Frequency stability is very good, and a bandreject filter notch depth to about -40 dB is generally available. It should be appreciated that the principal capability of the active filter network is to synthesize a complex-conjugate pole pair. This achievement, as described below, permits the realization of any mathematically definable lowpass approximation.

Kirchhoff's current law provides that the sum of the currents into any

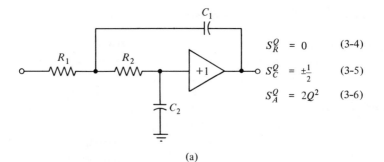

$$S_R^Q = 0 \qquad (3\text{-}4)$$

$$S_C^Q = \pm\frac{1}{2} \qquad (3\text{-}5)$$

$$S_A^Q = 2Q^2 \qquad (3\text{-}6)$$

(a)

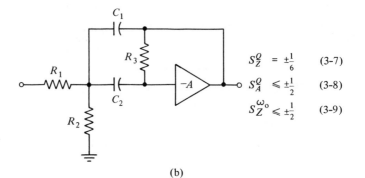

$$S_Z^Q = \pm\frac{1}{6} \qquad (3\text{-}7)$$

$$S_A^Q \leqslant \pm\frac{1}{2} \qquad (3\text{-}8)$$

$$S_Z^{\omega_o} \leqslant \pm\frac{1}{2} \qquad (3\text{-}9)$$

(b)

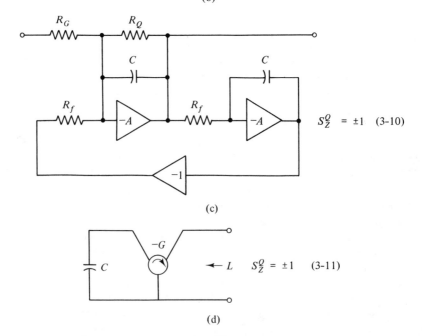

$$S_Z^Q = \pm 1 \qquad (3\text{-}10)$$

(c)

$$S_Z^Q = \pm 1 \qquad (3\text{-}11)$$

(d)

Figure 4-12 Recommended Active Filter Networks: (a) Unity Gain, (b) Multiple Feedback, (c) Biquad, (d) Gyrator

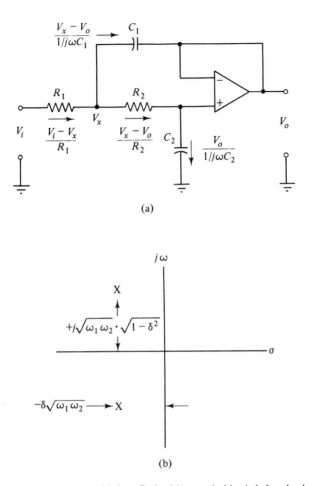

(a)

(b)

Figure 4-13 Unity-Gain Network Nodal Analysis

node is zero. A nodal analysis of the unity-gain lowpass network yields equations (4-10) through (4-14). It includes the assumption that current in C_2 is equal to current in R_2; the realization of this requires the use of a low-input-bias-current operational amplifier for accurate performance. The transfer function is obtained upon substituting for V_x in equation (4-10) its independent expression obtained from equation (4-11). Filter pole positions are defined by equation (4-14). Figure 4-13 shows these nodal equations and the complex-plane pole positions mathematically described by equation (4-14). This second-order network has two denominator roots (two poles) and is sometimes referred to as a resonator.

$$\frac{V_i - V_x}{R_1} = \frac{V_x - V_o}{1/j\omega C_1} + \frac{V_x - V_o}{R_2} \qquad (4\text{-}10)$$

$$\frac{V_x - V_o}{R_2} = \frac{V_o}{1/j\omega C_2} \qquad (4\text{-}11)$$

Rearranging,

$$V_x = V_o \cdot \frac{R_2 + 1/j\omega C_2}{1/j\omega C_2}$$

$$\frac{V_o}{V_i} = \frac{1}{\omega^2 R_1 R_2 C_1 C_2 + \omega C_2(R_1 + R_2) + 1} \qquad (4\text{-}12)$$

where
$$\omega_1 = \frac{1}{R_1 C_1} \quad \text{and} \quad \omega_2 = \frac{1}{R_2 C_2}$$

$$\delta = \frac{C_2}{2}(R_1 + R_2) \qquad (4\text{-}13)$$

$$s_{1,2} = -\delta\sqrt{\omega_1\omega_2} \pm j\sqrt{\omega_1\omega_2} \cdot \sqrt{1 - \delta^2} \qquad (4\text{-}14)$$

A recent technique using MOS technology has made possible the realization of multipole unity-gain network active filters in total integrated-circuit form without the requirement for external components. Small-value MOS capacitors are utilized with MOS switches in a switched-capacitor circuit for simulating large-value resistors under control of a multiphase clock. With reference to Figure 4-14, the rate f_s at which the capacitor is toggled determines its charging to V and discharging to V'. Consequently, the average current flow I from V to V' defines an equivalent resistor R that would provide the same average current shown by the identity of equation (4-15):

Figure 4-14 Switched Capacitor Unity-Gain Network

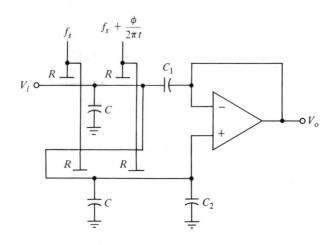

$$R = \frac{V - V'}{I} = 1/Cf_c \qquad (4\text{-}15)$$

The switching rate f_s is normally much higher than the signal frequencies of interest so that the time sampling of the signal can be ignored in a simplified analysis. Filter accuracy is primarily determined by the stability of the frequency of f_s and the accuracy of implementation of the monolithic MOS capacitor ratios. Reticon Corporation has pioneered in the manufacture of these devices, which offer considerable promise in terms of cost and space savings.

The most important parameter in the selection of operational amplifiers for active filter service is open-loop gain. The ratio of open-loop to closed-loop gain, or loop gain, must be 10^2 or greater for stable and well-behaved performance at the highest signal frequencies present. This is critical in the application of bandpass filters to insure a realization that accurately follows the design calculations. Amplifier input and output impedances are normally sufficiently close to the ideal infinite input and zero output values to be inconsequential at the impedance levels in active filter networks. Metal film resistors having a temperature coefficient of 50 ppm/°C are recommended for active filter design. Carbon composition resistors are entirely unsuitable, except where requirements are completely noncritical, because of their substantial temperature coefficient.

Selection of capacitor type is the most difficult decision because of many interacting factors. For most applications polystyrene capacitors are recommended because of their reliable -120 ppm/°C temperature coefficient and 0.05% capacitance retrace deviation with temperature cycling. Where capacitance values above 0.1 μF are required, however, polycarbonate capacitors are available in values to 1 μF with a ±50 ppm/°C temperature coefficient and 0.25% retrace. Mica capacitors are the most stable devices with ±50 ppm/°C tempco and 0.1% retrace, but practical capacitance availabilty is typically only 100 pF to 5000 pF. Mylar capacitors are available in values to 10 μF with 0.3% retrace, but their tempco averages 400 ppm/°C.

The choice of resistor and capacitor tolerance determines the accuracy of the filter implementation such as its cutoff frequency and passband flatness. Cost considerations normally dictate the choice of 1% tolerance resistors and 2% to 5% tolerance capacitors. However, it is usual practice to pair larger and smaller capacitor values to achieve required filter network values to within 1%, which results in filter parameters accurate to 1% or 2% with low tempco and retrace components. Filter response is typically displaced inversely to passive-component tolerance, such as lowering of cutoff frequency for component values on the high side of their tolerance band. For more critical realizations, such as high-Q bandpass filters, some provision for adjustment provides flexibility needed for an accurate implementation. Table 4-4 provides a compendium of the foregoing filter networks.

TABLE 4-4 *Summary of Previous Section*
ACTIVE FILTER NETWORK SELECTOR

FUNCTION	NETWORK	COMMENT
Lowpass	Unity gain	−100 dB/octave practical
Highpass	Unity gain	−100 dB/octave practical
Low-Q bandpass	Multiple feedback	Requires 1 op amp
High-Q bandpass	Biquad	Only choice $Q > 10$
Flat bandpass	Either bandpass network	Stagger tuned
Bandreject	Gyrator	−40-dB notch
	Lowpass + highpass	Very flat passband

4-3 ACTIVE FILTER REALIZATIONS

Table 4-5* provides the capacitor values in farads for unity-gain networks tabulated according to the number of filter poles. Higher-order filters are formed by a cascade of the second- and third-order networks shown in Figure 4-15,* each of which is different. For example, a sixth-order filter will have

TABLE 4-5A
UNITY-GAIN NETWORK CAPACITOR VALUES IN FARADS

	BUTTERWORTH			BESSEL		
Poles	C_1	C_2	C_3	C_1	C_2	C_3
2	1.414	0.707		0.907	0.680	
3	3.546	1.392	0.202	1.423	0.988	0.254
4	1.082	0.924		0.735	0.675	
	2.613	0.383		1.012	0.390	
5	1.753	1.354	0.421	1.009	0.871	0.309
	3.235	0.309		1.041	0.310	
6	1.035	0.966		0.635	0.610	
	1.414	0.707		0.723	0.484	
	3.863	0.259		1.073	0.256	
7	1.531	1.336	0.488	0.853	0.779	0.303
	1.604	0.624		0.725	0.415	
	4.493	0.223		1.098	0.216	
8	1.091	0.981		0.567	0.554	
	1.202	0.831		0.609	0.486	
	1.800	0.556		0.726	0.359	
	5.125	0.195		1.116	0.186	

1ˢᵗ Stage / *2ⁿᵈ Stage*

* *Table 4-5 and Figure 4-15 reprinted from *Electronics,* August 18, 1969, Copyright© McGraw-Hill, Inc., 1969. All rights reserved.

TABLE 4-5B
UNITY-GAIN NETWORK CAPACITOR VALUES IN FARADS

Poles	CHEBYSHEV 0.1-dB RIPPLE			CHEBYSHEV 1-dB RIPPLE		
	C_1	C_2	C_3	C_1	C_2	C_3
2	1.638	0.696		2.218	0.606	
3	6.653	1.825	0.135	16.180	2.567	0.064
4	1.901	1.241		3.125	1.269	
	4.592	0.241		7.546	0.149	
5	4.446	2.520	0.380	8.884	3.935	0.254
	6.810	0.158		0.115	0.093	
6	2.553	1.776		4.410	1.904	
	3.487	0.492		6.024	0.312	
	9.531	0.111		16.460	0.064	
7	5.175	3.322	0.569	10.290	5.382	0.401
	4.546	0.333		7.941	0.199	
	1.273	0.082		22.250	0.047	
8	3.270	2.323		5.756	2.538	
	3.857	0.689		6.792	0.443	
	5.773	0.240		10.150	0.139	
	16.440	0.063		28.940	0.036	

six different capacitor values and not consist of a cascade of identical two-pole or three-pole networks. Figures 4-16 and 4-17 illustrate the design procedure with 1-kHz-cutoff two-pole Butterworth lowpass and highpass filters including the frequency and impedance scaling steps. The three-pole filter design procedure is identical with observation of the appropriate network capacitor locations but should be driven from a low driving-point impedance such as an operational amplifier. A design guide for unity-gain active filters is summarized in the following steps.

1. Select an appropriate filter approximation and number of poles required to provide the necessary response from the curves of Figures 4-1 through 4-9.
2. Choose the filter network appropriate for the required realization from Figure 4-15 and perform the necessary component frequency and impedance scaling.
3. Implement the filter components by selecting 1% standard-value resistors and then pairing a larger and smaller capacitor to realize each capacitor value to within 1%.

A multiple-feedback bandpass filter (MFBF) is shown in Figure 4-18 with a center frequency of 1 Hz and a Q of 10. Equations (4-16) through (4-19) derive the component values for this filter. Normally, a standard capacitor

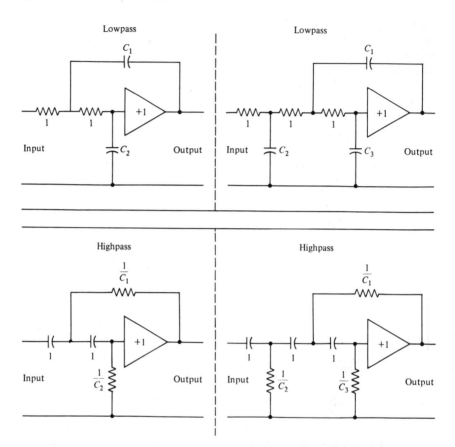

Figure 4-15 Two- and Three-Pole Unity-Gain Networks

value C is chosen in a range that results in reasonable resistor values with components selected to 1% tolerance. It should be noted that this circuit produces a signal inversion. When higher-Q active bandpass filtering is required, the biquad network must be considered. Although its mechanization does require three operational amplifiers, the biquad provides the capability to independently set filter center frequency f_0, Q, and gain A_{f_0} at the center frequency. A practical design approach is to fix the frequency-determining resistors R_{f_0} shown in Figure 4-19 at a standard value, and then calculate the other component values as presented by equations (4-20)–(4-22) and tabulated in Table 4-6 for representative instrumentation frequencies.

Most instrumentation systems involve amplitude measurements of transducer outputs, and it is normally of interest to maintain amplitude flatness in the signal passband. In the case of bandpass filtering using the previous single-tuned networks, the amplitude response rolls off immediately on both sides of the center frequency. Bandpass signals having an extended spectral occupancy, therefore, should be filtered by a flat-passband bandpass filter. A stagger-tuning scheme for multiple-cascaded single-tuned bandpass filters can produce a flat passband with the additonal benefit of increased skirt selectivity. Table 4-6 presents stagger-tuning parameters for a maximally flat passband in terms of the number of single-tuned networks employed, their individual center frequencies f_r and -3-dB bandwidths Δf_r, and the overall bandpass filter center frequency f_0 and -3-dB bandwidth Δf. Passband

Figure 4-16 Butterworth Unity-Gain Lowpass Filter for $f_C = 1\,KHz$

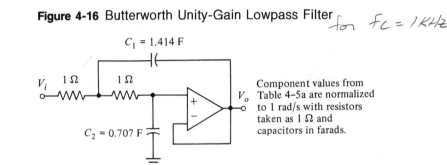

$C_1 = 1.414$ F

V_i 1 Ω 1 Ω

V_o

$C_2 = 0.707$ F

Component values from Table 4-5a are normalized to 1 rad/s with resistors taken as 1 Ω and capacitors in farads.

(a)

$$\frac{1.414\ \text{F}}{(2\pi)(1\ \text{kHz})} = 225\ \mu\text{F}$$

$$\frac{0.707\ \text{F}}{(2\pi)(1\ \text{kHz})} = 112.5\ \mu\text{F}$$

The filter is then frequency-scaled by dividing the capacitor values from the table by the cutoff frequency in radians ($2\pi \times 1$ kHz).

(b)

$$\frac{225\ \mu\text{F}}{10\ \text{k}} = 0.0225\ \mu\text{F}$$

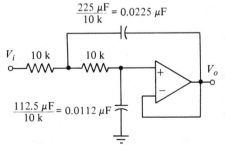

V_i 10 k 10 k

V_o

$$\frac{112.5\ \mu\text{F}}{10\ \text{k}} = 0.0112\ \mu\text{F}$$

The filter is finally impedance-scaled by multiplying the resistor values by a convenient value (10 k) and dividing the capacitor values by the same value.

(c)

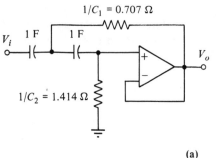

$1/C_1 = 0.707\ \Omega$

V_i 1 F 1 F

$1/C_2 = 1.414\ \Omega$

V_o

Component values from Table 4-5a are normalized to 1 rad/s with capacitors taken as 1 F and resistors the (inverse) *reciprocal?* capacitor values from the table in ohms.

(a)

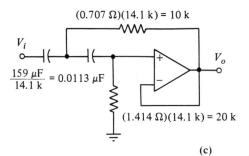

$$\frac{1\ \text{F}}{(2\pi)(1\ \text{kHz})} = 159\ \mu\text{F}$$

The filter is then frequency-scaled by dividing the capacitor values by the cutoff frequency in radians of value ($2\pi \times$ 1 kHz).

(b)

$(0.707\ \Omega)(14.1\ \text{k}) = 10\ \text{k}$

V_i

$$\frac{159\ \mu\text{F}}{14.1\ \text{k}} = 0.0113\ \mu\text{F}$$

V_o

$(1.414\ \Omega)(14.1\ \text{k}) = 20\ \text{k}$

The filter is finally impedance-scaled by multiplying the resistor values by a convenient value (14.1 k) and dividing the capacitor values by the same value.

(c)

for $f_c = 1\,KHz$

Figure 4-17 Butterworth Unity-Gain Highpass Filter

flatness and skirt selectivity both improve, of course, as the number of cascaded single-tuned networks increases and the overall Δf decreases.

Consider, for example, a bandpass filter requirement centered at an f_0 of 1 kHz with a maximally flat Δf bandwidth of 200 Hz. This $Q = 5$ filter is also to achieve -35-dB attenuation ± 1 octave on both sides of the center frequency f_0. Two cascaded and stagger-tuned MFBF networks are able to meet these specifications requiring only two operational amplifiers for their implementation. The individual MFBF networks are designed according to the example associated with Figure 4-18, but employing the tuning parameters obtained from Table 4-7. The filter circuit is shown by Figure 4-20 with 0.1-μF capacitors. In the event that final minor tuning adjustments are required, each network center frequency is determined by R_2, Q by R_3, and gain by R_1. A penalty of the stagger-tuned method is a gain loss that results from the

$C_1 = C_2$

$R_3 = 2R_1$

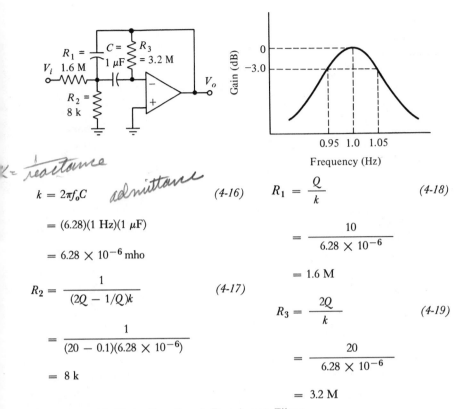

$k = \text{reactance}$

$k = 2\pi f_o C \quad \text{admittance}$ (4-16)

$= (6.28)(1 \text{ Hz})(1\ \mu\text{F})$

$= 6.28 \times 10^{-6} \text{ mho}$

$R_2 = \dfrac{1}{(2Q - 1/Q)k}$ (4-17)

$= \dfrac{1}{(20 - 0.1)(6.28 \times 10^{-6})}$

$= 8\ \text{k}$

$R_1 = \dfrac{Q}{k}$ (4-18)

$= \dfrac{10}{6.28 \times 10^{-6}}$

$= 1.6\ \text{M}$

$R_3 = \dfrac{2Q}{k}$ (4-19)

$= \dfrac{20}{6.28 \times 10^{-6}}$

$= 3.2\ \text{M}$

Figure 4-18 Multiple-Feedback Bandpass Filter

algebraic addition of the skirts of each network. However, this loss may be calculated and compensated for on a per-network basis as shown in the example calculations that follow. The overall filter response achieved is described by Figure 4-21 and has a $+0.3$-dB amplitude response at ±70 Hz (70% bandwidth) on either side of the 0-dB f_0.

Figure 4-19 Biquad Bandpass Filter

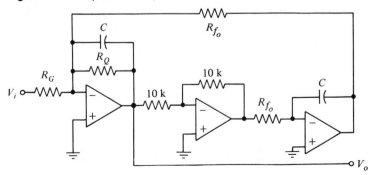

TABLE 4-6
BIQUAD COMPONENT VALUES FOR $R_{f_o} = 10$ k

f_o	$C = \dfrac{1}{(2\pi f_o R_{f_o})}$ (4-20)	Q	$R_q = \dfrac{Q}{2\pi f_o C}$ (4-21)	A_{f_o}	$R_g = \dfrac{R_q}{A_{f_o}}$ (4-22)
10 Hz	1.6 μF	10	100 k	$Q/100$	1 M
100 Hz	0.16 μF	50	500 k	$Q/50$	500 k
1 kHz	0.016 μF	100	1 M	$Q/10$	100 k
10 kHz	0.0016 μF	200	2 M	Q	10 k

TABLE 4-7
STAGGER-TUNING PARAMETERS

SINGLE-TUNED FILTERS	Δf_r	f_r
2	0.71 Δf	$f_o + 0.35\ \Delta f$
	0.71 Δf	$f_o - 0.35\ \Delta f$
3	0.5 Δf	$f_o + 0.43\ \Delta f$
	0.5 Δf	$f_o - 0.43\ \Delta f$
	1.0 Δf	f_o
4	0.38 Δf	$f_o + 0.46\ \Delta f$
	0.38 Δf	$f_o - 0.46\ \Delta f$
	0.93 Δf	$f_o + 0.19\ \Delta f$
	0.93 Δf	$f_o - 0.19\ \Delta f$

Figure 4-20 Stagger-Tuned Multiple-Feedback Bandpass Filter

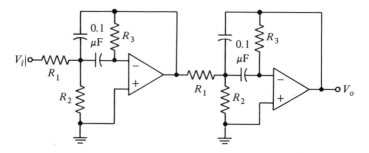

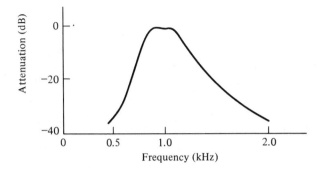

Figure 4-21 Stagger-Tuned $Q=5$ Bandpass Response

$$\text{gain loss}_r = \frac{A_r}{\sqrt{A_r^2 + B_r^2}} \qquad (4\text{-}23)$$

$$A_r = \frac{(2\pi f_r)(2\pi f_g)}{Q_r} \qquad (4\text{-}24)$$

$$B_r = (2\pi f_r)^2 - (2\pi f_g)^2 \qquad (4\text{-}25)$$

$$f_g = \sqrt{f_u \cdot f_L} \qquad (4\text{-}26)$$

$$= \sqrt{(1100 \text{ Hz})(900 \text{ Hz})}$$

$$= 995 \text{ Hz}$$

First Section:

$$\Delta f_{r_1} = 0.71 \, \Delta f$$

$$= (0.71)(200 \text{ Hz})$$

$$= 141 \text{ Hz}$$

$$f_{r_1} = f_o + 0.35 \, \Delta f$$

$$= 1 \text{ kHz} + (0.35)(200 \text{ Hz})$$

$$= 1.07 \text{ kHz}$$

Second Section:

$$\Delta f_{r_2} = 0.71 \, \Delta f$$

$$= (0.71)(200 \text{ Hz})$$

$$= 141 \text{ Hz}$$

$$f_{r_2} = f_o - 0.35 \, \Delta f$$

$$= 1 \text{ kHz} - (0.35)(200 \text{ Hz})$$

$$= 930 \text{ Hz}$$

First Section:

$$Q_1 = \frac{f_{r_1}}{\Delta f_{r_1}} \qquad (4\text{-}27)$$

$$= \frac{1.07 \text{ kHz}}{141 \text{ Hz}}$$

$$= 7.6$$

$$k_1 = 2\pi f_{r_1} C \qquad (4\text{-}16)$$

$$= (2\pi)(1.07 \text{ kHz})(0.1 \ \mu\text{F})$$

$$= 6.72 \times 10^{-4} \text{ mho}$$

$$A_1 = \frac{(2\pi f_{r_1})(2\pi f_g)}{Q_1} \qquad (4\text{-}24)$$

$$= \frac{(2\pi \cdot 1.07 \text{ kHz})(2\pi \cdot 995 \text{ Hz})}{7.6}$$

$$= 5.53 \times 10^6$$

$$B_1 = (2\pi f_{r_1})^2 - (2\pi f_g)^2 \qquad (4\text{-}25)$$

$$= (2\pi \cdot 1.07 \text{ kHz})^2 - (2\pi \cdot 995 \text{ Hz})^2$$

$$= 6.15 \times 10^6$$

$$\text{gain loss}_1 = \frac{A_1}{\sqrt{A_1^2 + B_1^2}} \qquad (4\text{-}23)$$

$$= \frac{5.53 \times 10^6}{\sqrt{(5.53 \times 10^6)^2 + (6.15 \times 10^6)^2}}$$

$$= 0.669$$

$$R_1 = \frac{Q_1 \cdot \text{gain loss}_1}{K_1} \qquad (4\text{-}28)$$

$$= \frac{(7.6)(0.669)}{6.72 \times 10^{-4} \text{ mho}}$$

$$= 7.55 \text{ k}$$

Second Section:

$$Q_2 = \frac{f_{r_2}}{\Delta f_{r_2}} \qquad (4\text{-}27)$$

$$= \frac{930 \text{ Hz}}{141 \text{ Hz}}$$

$$= 6.6$$

$$k_2 = 2\pi f_{r_2} C \qquad (4\text{-}16)$$

$$= (2\pi)(930 \text{ Hz})(0.1 \ \mu\text{F})$$

$$= 5.85 \times 10^{-4} \text{ mho}$$

$$A_2 = \frac{(2\pi f_{r_2})(2\pi f_g)}{Q_2} \qquad (4\text{-}24)$$

$$= \frac{(2\pi \cdot 930 \text{ Hz})(2\pi \cdot 995 \text{ Hz})}{6.6}$$

$$= 5.53 \times 10^6$$

$$B_2 = (2\pi f_{r_2})^2 - (2\pi f_g)^2 \qquad (4\text{-}25)$$

$$= (2\pi \cdot 930 \text{ Hz})^2 - (2\pi \cdot 995 \text{ Hz})^2$$

$$= -4.9 \times 10^6$$

$$\text{gain loss}_2 = \frac{A_2}{\sqrt{A_2^2 + B_2^2}} \qquad (4\text{-}23)$$

$$= \frac{5.53 \times 10^6}{\sqrt{(5.53 \times 10^6)^2 + (-4.9 \times 10^6)^2}}$$

$$= 0.75$$

$$R_1 = \frac{Q_2 \cdot \text{gain loss}_2}{K_2} \qquad (4\text{-}28)$$

$$= \frac{(6.6)(0.75)}{5.85 \times 10^{-4} \text{ mho}}$$

$$= 8.47 \text{ k}$$

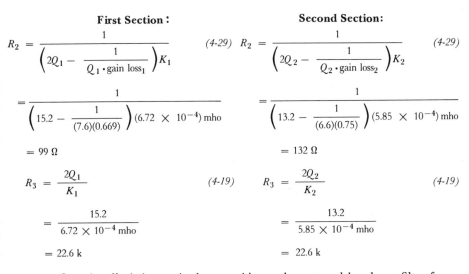

First Section :

$$R_2 = \cfrac{1}{\left(2Q_1 - \cfrac{1}{Q_1 \cdot \text{gain loss}_1}\right)K_1} \quad \text{(4-29)}$$

$$= \cfrac{1}{\left(15.2 - \cfrac{1}{(7.6)(0.669)}\right)(6.72 \times 10^{-4})\,\text{mho}}$$

$$= 99\ \Omega$$

$$R_3 = \frac{2Q_1}{K_1} \quad \text{(4-19)}$$

$$= \frac{15.2}{6.72 \times 10^{-4}\,\text{mho}}$$

$$= 22.6\ \text{k}$$

Second Section:

$$R_2 = \cfrac{1}{\left(2Q_2 - \cfrac{1}{Q_2 \cdot \text{gain loss}_2}\right)K_2} \quad \text{(4-29)}$$

$$= \cfrac{1}{\left(13.2 - \cfrac{1}{(6.6)(0.75)}\right)(5.85 \times 10^{-4})\,\text{mho}}$$

$$= 132\ \Omega$$

$$R_3 = \frac{2Q_2}{K_2} \quad \text{(4-19)}$$

$$= \frac{13.2}{5.85 \times 10^{-4}\,\text{mho}}$$

$$= 22.6\ \text{k}$$

Occasionally it is required to provide a voltage-tuned bandpass filter for signal-tracking purposes. The biquad mechanization is especially suitable for this purpose, because its Q and passband gain will remain fixed when its center frequency f_o is made to vary by adjustment of the R_{f_o} resistors. In practice, linear four-quadrant analog multipliers, such as the 1% error Analog Devices AD532 devices, may be applied to vary the conductance of these center-frequency-determining resistors over a practical two-decade range with excellent tracking. Figure 4-21 shows the amplitude response for two stagger-tuned biquad networks providing a $Q = 5$ maximally flat passband response over a tuning range from 10 Hz to 1 kHz achieved with approximately $\frac{1}{2}$ dB amplitude variation over this range. The filter is designed at the maximum center frequency of 1 kHz, and then the conductance of R_{f_o} is decreased to reduce f_0 from 1 kHz to 10 Hz by decreasing the multiplier tuning voltage from 10 V to 100 mV. The circuit for this voltage-tunable filter including its component values, as shown in Figure 4-22, uses the previous stagger-tuned bandpass filter calculations including equations (4-23) through (4-29).

The gyrator bandreject filter realization is described by Figure 4-23 and equation (4-30). Frequency stability is very good, but the realization of Q values greater than about 5 requires an amplifier-loop gain of 10^3 or greater at the notch frequency f_c. Also, at higher Q values signal-input amplitude may have to be attenuated to preserve linearity. However, a notch depth to -40 dB is available with this circuit. An alternate and more flexible method of forming a bandreject filter is shown by Figure 4-24, in which the parallel outputs of unity-gain lowpass and highpass networks are summed. The Q provided by this bandreject filter is dependent on the order of the unity-gain networks, but an advantage is the excellent passband flatness maintained above and below the cutoff frequencies. The two-operational-amplifier Butterworth implementation suggested by Figure 4-24 has a notch depth of −18 dB and Q of 1 at 60 Hz.

Figure 4-22 Voltage-Tuned Biquad Bandpass Filter

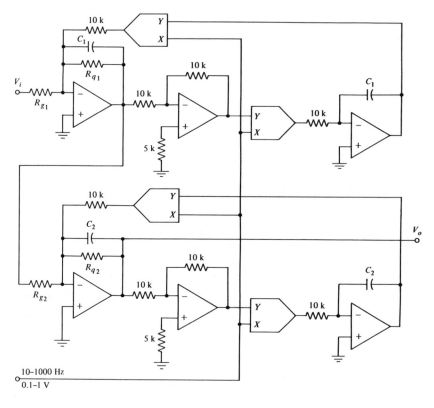

Gain Loss$_1$ = 0.669
R_{fo} = 10 k
fo_1 = 1070 Hz
Q_1 = 7.6

$$C_1 = \frac{1}{(2\pi)(10\text{ k})(1.07\text{ kHz})} = 0.0149\ \mu F$$

$$R_{q1} = \frac{Q_1}{2\pi fo_1 C_1} = 81.3\text{ k}$$

$$A_{fo_1} = \frac{1}{\text{gain loss}_1} = 1.5$$

$$R_{g1} = \frac{R_{q1}}{A_{fo_1}} = 54.2\text{ k}$$

Gain Loss$_2$ = 0.75
R_{fo} = 10 k
fo_2 = 930 k
Q_2 = 6.6

$$C_2 = \frac{1}{(2\pi)(10\text{ k})(0.93\text{ kHz})} = 0.0171\ \mu F$$

$$R_{q2} = \frac{Q_2}{2\pi fo_2 C_2} = 61.5\text{ k}$$

$$A_{fo_2} = \frac{1}{\text{gain loss}_2} = 1.33$$

$$R_{g2} = \frac{R_{q2}}{A_{fo_2}} = 46.3\text{ k}$$

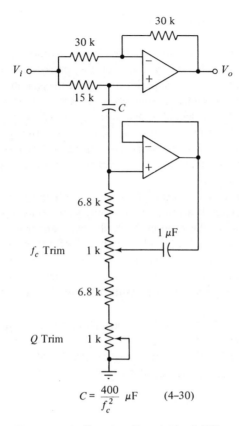

$$C = \frac{400}{f_c^2} \ \mu F \qquad (4\text{--}30)$$

Figure 4-23 Gyrator Bandreject Filter

Figure 4-24 Lowpass Plus Highpass Bandreject Filter

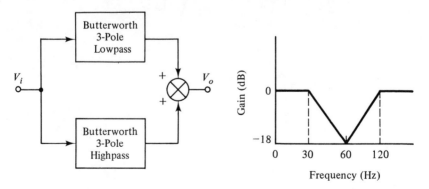

4-4 FILTER ERROR ANALYSIS

Active filter approximations for bandlimiting instrumentation signals can be realized inexpensively by the methods of the preceding sections. However, when a filter characteristic is superimposed on a measurement signal, its gain and phase errors are added to the error budget of the signal-conditioning channel. It is therefore essential to determine the filter component error, which depends upon the filter type, its number of poles, and the type of signal. Since instrumentation-filter requirements are typically lowpass with either dc, sinusoidal, or complex harmonic signal types, an error analysis is performed for these conditions.

An example of why linear phase delay with frequency is important is shown by Figure 4-25. If a complex harmonic signal consists of a fundamental and third harmonic in the ratio of 2 : 1, the resultant waveform described by Figure 4-25(a) has equal amplitude maximums of 1.0 units as shown. If this signal is passed through a six-pole lowpass Butterworth filter whose passband includes both fundamental and harmonic frequencies, the distorted waveform of Figure 4-25(b) is produced as a result of its nonlinear phase characteristic.

Figure 4-25 Filtered Complex Waveform Phase Nonlinearity: (a) Sum of Fundamental and Third Harmonic in 2:1 Ratio, (b) Sum of Fundamental and Third Harmonic Following Six-Pole Lowpass Butterworth Filter *(Courtesy Electronic Instrumentation and Technology)*

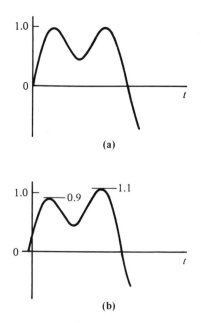

The differences in time delay between signal components at different frequencies encountered during filtering cause their displacement and the amplitude alteration illustrated by Figure 4-25(b). The effect of filter phase deviation from a linear phase delay, described by equation (4-31), is the average phase error affecting filter amplitude accuracy, where $\Delta\phi$ is deviation from linear phase in degrees. Filter passband phase error and gain error may then be combined by equation (4-32) in an rms fashion, owing to their independence, to provide amplitude error at specific passband frequencies. Table 4-8 presents a tabulation of these results for three-pole lowpass Butterworth, Bessel, one-pole RC, and Besselworth filters, including their average amplitude error, from equation (4-33), up to indicated fractions of the filter passbands.

An evident conclusion from the average-error column of Table 4-8 is that for complex harmonic waveforms, such as encountered with video and biomedical signals, the filter error contribution can be minimized by insuring that the filter cutoff frequency is greater than the signal spectral occupancy requirements defined in Table 1-2. For example, the three-pole Bessel filter exhibits a moderate 0.75% error to $0.4f_c$ for this relatively difficult signal type. An error analysis of a five-pole Bessel lowpass yields 1.75% error at the same $0.4f_c$ passband fraction. Consequently, to prevent excessive filter component error with complex harmonic signal types, higher-order Bessel filters should be avoided and Butterworth filters should not be used. Sharper rolloff requirements for complex harmonic signals necessitate the use of proprietary amplitude-error compensated filters represented by the four-pole Besselworth device offered by Electronic Instrumentation and Technology. This filter has an amplitude response like that of a four-pole Butterworth lowpass, with its phase deviation compensated such that the amplitude error is essentially attributable to its gain error. The average passband error of this device is a nominal 0.2% up to $0.6f_c$.

DC and sinusoidal signals do not result in the same filter amplitude error magnitudes, since their single spectral term is insensitive to phase-deviation anomalies. Filter gain error is therefore the predominant error source and is minimized with Butterworth filters because of their flat passband amplitude characteristic. Consequently, three- to five-pole Butterworth lowpass filters exhibit an average passband error of about 0.2% to $0.6f_c$ and are recommended for dc and sinusoidal signals. For comparison purposes a one-pole RC passive filter is represented in Table 4-8; it has an average passband error comparable to that of the three-pole Butterworth filter for complex harmonic signals. However, its gain error exhibits a faster buildup than that of the Butterworth filter. Consequently, its passband utilization must be limited to insure a low filter component error even with dc and sinusoidal signals. Utilization to $0.1f_c$ provides 0.25% error, which rises to 1.25% at $0.2f_c$; these are average errors determined from gain error only with reference to 0 Hz. Table 4-9 summarizes these filter application recommendations.

TABLE 4-8
LOWPASS FILTER ERRORS FOR COMPLEX HARMONIC SIGNALS

FREQUENCY $\frac{f}{f_c}$	GAIN ERROR (0 Hz REFERENCE)				Δφ DEVIATION (DEGREES)				AMPLITUDE ERROR (%FS)				AVERAGE ERROR (%FS)			
	3-Pole Butterworth	3-Pole Bessel	4-Pole Besselworth	1-Pole RC	3-Pole Butterworth	3-Pole Bessel	4-Pole Besselworth	1-Pole RC	3-Pole Butterworth	3-Pole Bessel	4-Pole Besselworth	1-Pole RC	3-Pole Butterworth	3-Pole Bessel	4-Pole Besselworth	1-Pole RC
0.0	0	0	0	0	0	0	0	0	0	0	0	0	0	0	0	0
0.1	0	0	0	0.005	2	0	0.2	1.5	2	0	0	1				
0.2	0	0	0	0.02	5	0	0.3	3.0	4	0	0	3	3	0	0	2
0.3	0	0.01	0	0.04	7	0	0.3	3.5	6	1	0	5				
0.4	0	0.02	0	0.07	9	0	0.2	4.0	8	2	0	8	5	0.75	0	4
0.5	0	0.04	0	0.10	10	2	0.1	4.5	10	4	0	10				
0.6	0.01	0.06	0.01	0.14	11	4	0.2	4.0	10	8	1	15	7	2.5	0.2	7
0.7	0.05	0.08	0.04	0.18	10	7	0.1	3.5	10	11	4	20				
0.8	0.12	0.11	0.11	0.22	8	11	0.5	3.0	13	15	11	23	8	5	2	11
0.9	0.20	0.16	0.18	0.26	4	15	1.1	1.5	20	20	18	26				
1.0	0.29	0.23	0.29	0.29	0	20	0	0	29	29	29	29	11	9	6	14

TABLE 4-9
LOWPASS FILTER APPLICATION RECOMMENDATIONS

SIGNAL	FILTER	ERROR (%FS)
DC and sinusoidal	3-5-Pole Butterworth	0.2% to $0.6f_c$
	1-Pole RC	0.25% to $0.1f_c$
Complex harmonic	4-Pole Besselworth	0.2% to $0.6f_c$
	3-Pole Bessel	0.75% to $0.4f_c$

$$\text{phase error} = \sqrt{\frac{1 - \cos \Delta\phi}{2}} \times 100\% \qquad (4\text{-}31)$$

$$\text{amplitude error} = \sqrt{(\text{gain error})^2 + (\text{phase error})^2} \qquad (4\text{-}32)$$

$$\text{average error} = \frac{0.1}{f/f_c} \cdot \sum_0^{f/f_c} (\text{amplitude error}) \qquad (4\text{-}33)$$

REFERENCES

1. R. P. SALLEN and E. L. KEY, "A Practical Method of Designing RC Active Filters," *IRE Transactions on Circuit Theory*, Vol. CT-2, March 1955.

2. P. R. GEFFE, "Toward High Stability in Active Filters," *IEEE Spectrum*, Vol. 7, May 1970.

3. L. C. THOMAS, "The Biquad, Part 1—Some Practical Design Considerations," *IEEE Circuit Theory Transactions*, Vol. CT-18, May 1971.

4. S. K. MITRA, "Synthesizing Active Filters," *IEEE Spectrum*, Vol. 6, January 1969.

5. R. BRANDT, "Active Resonators Save Steps in Designing Active Filters," *Electronics*, April 24, 1972.

6. B. ZEINES, *Introduction to Network Analysis*, Prentice-Hall, Englewood Cliffs, N.J., 1967.

7. C. MITRA, *Analysis and Synthesis of Linear Active Networks*, John Wiley, New York, 1969.

8. J. W. CRAIG, *Design of Lossy Filters*, MIT Press, Cambridge, Mass., 1970.

9. R. W. DANIELS, *Approximation Methods for Electronic Filter Design*, McGraw-Hill, New York, 1974.

10. D. E. JOHNSON, *Introduction to Filter Theory*, Prentice-Hall, Englewood Cliffs, N.J., 1976.

11. D. E. JOHNSON and J. L. HILBURN, *Rapid Practical Designs of Active Filters*, John Wiley, New York, 1975.

12. A. B. WILLIAMS, *Active Filter Design*, Airtech House, Dedham, Mass., 1975.

13. J. T. MAY, *Antialiasing Filters in Digital Filter Data Collection Systems: A New Approach*, Electronic Instrumentation and Technology, Sterling, Va. 1979.

PROBLEMS

4-1 Design a seven-pole Butterworth lowpass filter having a 1-Hz cutoff frequency. Use unity-gain networks and 1-M resistors. Sketch the circuit including its component values plus a plot of its frequency response using Figure 4-2 and Table 4-5.

4-2. Implement the bandreject filter of Figure 4-24 using unity-gain networks, two operational amplifiers, and 100-k resistors for impedance scaling with the aid of Table 4-5.

4-3. Design a 0.1-Hz-cutoff three-pole 1-dB ripple Chebyshev lowpass filter employing 10-M resistors for impedance scaling with the aid of Table 4-5. Provide a plot of its frequency response to 1 Hz and compensation for amplifier input-bias-current offset voltage.

4-4. Select the component values for a biquad bandpass filter with the aid of Table 4-6, and show the circuit for a 100-Hz center frequency, Q of 100, and unity gain.

4-5. The FM quadracast system proposed by General Electric includes a signal between 60 kHz and 90 kHz at the detector output, which must be selectively filtered for further processing. Design a cascade of three stagger-tuned multiple-feedback bandpass filters that preserves passband flatness while achieving the minimum stopband attenuation shown at 53 kHz. Use 0.001-μF capacitors, show all calculations plus the final circuit, and include a plot of the frequency response including attenuation at 53 kHz and 90 kHz.

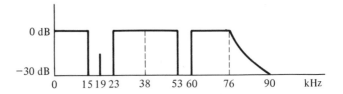

4-6. Using the methods of Section 4-4, determine the average error between dc and f_c for a five-pole Bessel lowpass filter with the aid of Figures 4-8 and 4-9, and determine the fraction of passband signal occupancy that both maximizes filter utilization and minimizes its error contribution. Then compare three- and five-pole Bessel filters at 0.75% average error in terms of stopband attenuation.

measurement sensors

5-0 INTRODUCTION

Modern technology leans heavily on the art and science of measurement. The operation of industrial processes, automated manufacturing, and more esoteric applications such as remote geophysical sensing would be impossible without accurate measuring devices. Those responsible for transducer-aided measurements must understand the energy translations involved in converting the quantity of interest into a useful readout and must appreciate their inexactness and the need for reference standards.

This chapter surveys electrical transducer techniques and devices that are useful for both industrial and laboratory measurements. It defines key transducer characteristics and discusses a variety of devices used for measurements in sensor-based data acquisition systems. Transducers for temperature, pressure, and flow are presented as well as those for displacement, strain, and vibration. Photometric and radiometric devices are described, and analytical instruments such as chemical analyzers introduced. In practice the words *sensor* and *transducer* are used interchangeably, although the former more accurately describes the device and the latter the principle involved.

5-1 BASIC PRINCIPLES

The function of a transducer is to sense the presence, magnitude, and rate of change of some measurand and to provide an electrical output that, when appropriately conditioned, furnishes a measurement signal to the desired accuracy. Transducers are typically classified by the electrical principle involved in their operation. The actual functioning of a particular transducer is of less interest here than the manner in which electrical outputs are developed. Unquestionably the science of measurement, or metrology, was significantly influenced by the development of electronic techniques that provided improved accuracy and the convenience of remote measurements.

A transducer is a device that transfers energy between two systems, as in the conversion of thermal into electrical energy by the Seebeck-effect thermocouple. Figure 5-1 describes a basic transducer and its relationship to the measurand quantity. Four transducer parameters frequently of interest are listed below. The *resolution* of a measurement is determined by the number of significant figures to which it is expressed, and the *precision* of the measurement by the variation of these figures over some measurement interval. The *accuracy* and *error* of a measurement refer to its closeness to the true value, the former to the degree of approach and the latter to the degree of departure. Since no measurement can be made with perfect accuracy, we need to determine what errors have entered into the measurement and what methods are best for reducing them. Errors may generally be classified as *systematic* or *bias* errors, such as produced by thermal offset voltages, or *random* errors, which are due to causes that cannot be directly established, such as induced noise. Measurement signal-conditioning operations presented in Chapter 6 provide an analysis of and methods for minimizing these errors.

Reference standards are involved in all measurements, either directly or indirectly, the more immediately as the requirement for accuracy increases. For example, highly accurate measurements are generally obtained by the comparison method, such as by using a voltage potentiometer employing a standard cell as defined by the National Bureau of Standards. A more common method is the periodic comparison and calibration of a transducer against a transfer standard by means of a correction procedure. At the minimum, usually, an initial calibration procedure is performed by the manufacturer.

accuracy: closeness with which a measurement approaches the true value of a measurand, expressed usually as a percent of full-scale output.

precision: an expression of the repeatability of measurements determined from the number of significant figures available.

Figure 5-1 Transducer Representation

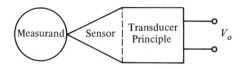

resolution: an expression of the smallest significant number to which a measurement can be determined.

error: the deviation of a measurement from the true value of a measurand, expressed usually as a percent of the full-scale output.

5-2 TEMPERATURE TRANSDUCERS

Temperature sensing is often required in industrial and laboratory measurements. The specific application and temperature measurement range largely dictate the choice of sensor. However, thermocouples, resistance-thermometer devices (RTD), semiconductor sensors, and pyrometers are dominant. Table 5-1 gives the ranges of temperature-measurement devices.

Thermocouple transducers are the most widely used temperature devices because of their ruggedness and broad temperature range. Two dissimilar metals are used in this Seebeck-effect temperature-to-emf generator with repeatable results as illustrated in Figure 5-2. Proper operation does require the use of a reference junction, or the electrical equivalent, in series with the measurement junction in order to polarize the direction of current flow and maximize the emf at the measurement junction. Various dissimilar-metal combinations are employed, depending upon the operating environment and the range of operating temperatures. The more frequently used combinations are listed in Table 5-2. Wide-range temperature measurements usually require linearization of the transfer characteristic.

When accurate thermocouple measurements are required, it is necessary to reference both legs to copper wire at an ice-point reference junction. Since a reference-junction temperature change influences the output signal and since ice baths are inconvenient, practical alternate methods are usually employed. The electrical bridge cold-junction compensator incorporates a temperature-sensitive resistance element that is one leg of the bridge network and thermally

TABLE 5-1
TEMPERATURE-SENSOR RANGES

TYPE	RANGE (°C)
Glass stem	−50 to +600
Bimetallic	−50 to +500
Filled element	−50 to +300
Semiconductor	−100 to +100
RTD	−100 to +300
Pyrometer	+100 to 5000
Thermocouple	−250 to 2000

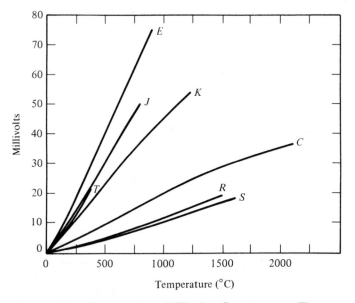

Figure 5-2 Temperature-Millivolt Graph for Thermocouples (*Courtesy Omega Engineering, Inc., an Omega Group Company*

integrated with the cold junction as shown in Figure 5-3. As the ambient temperature surrounding the cold junction varies, the thermally generated error voltage is compensated by an equal but opposite voltage in series with it. A *thermopile* is a series-connected arrangement of thermocouples providing an output equal to that of each thermocouple multiplied by the number of identical thermocouples used.

　　Resistance-thermometer devices, which provide greater resolution and repeatability than thermocouples, operate on the principle of electrical resistance change as a function of temperature. The platinum resistance thermometer is frequently used for industrial applications and offers good

TABLE 5-2
THERMOCOUPLE COMPARISON DATA

TYPE	*ELEMENTS, +/−*	*mV/°C*
E	Chromel/constantan	0.044
J	Iron/constantan	0.033
K	Chromel/alumel	0.020
R&S	Pt-Rh/platinum	0.010
T	Copper/constantan	0.040
C	Tungsten/rhenium	0.012

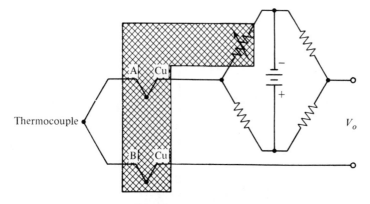

Figure 5-3 Electrical Reference Junction

accuracy and mechanical and electrical stability. *Thermistors* are fabricated from a sintered mixture of metal alloys forming a ceramic that exhibits a large negative temperature coefficient. Metal film resistors have an extended and more linear range than thermistors, but thermistors have about ten times the sensitivity. The signal conditioning usually required is conversion of the resistance change into a voltage change and possibly linearization. Figure 5-4 presents the temperature-to-resistance characteristic of popular RTD devices.

The *optical pyrometer* is useful for temperature measurement when mechanical-sensor contact with the process is not feasible, but a direct view is available. Measurements are limited to energy within the spectral response of the sensor used; a radiometric match between a calibrated reference source and the process provides a current output corresponding to a specific temperature. Automatic pyrometers employ a servo loop to provide this balance, as shown in Figure 5-5. Operation to 5000°C is typical.

Temperature measurement using forward-biased semiconductor devices is a new technique capable of accuracy to 0.1°C over a useful range of about ±100°C. Fortunately, many required temperature measurements fall within

RANGE (°C)	ERROR (%FS)	APPLICATION
0 to 800	0.5	High output
−250 to 700	0.75	Reducing atmospheres
−250 to 1200	0.75	Oxidizing atmospheres
0 to 1400	0.25	Corrosive atmospheres
−250 to 350	1.0	Moist atmospheres
0 to 2000	0.5	High temperature

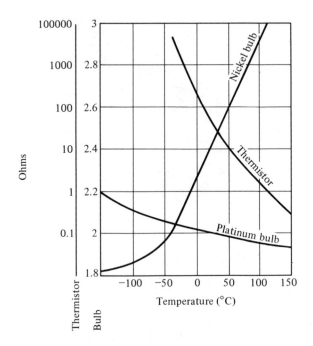

Figure 5-4 RTD Devices

this range. The negative temperature coefficient of the bipolar transistor base-to-emitter forward voltage drop varies by 2.5 mV per degree centigrade and can be made very linear by means of a constant-current supply connected as shown in Figure 5-6. This general type of temperature transducer is represented by National LX 5600 Series devices.

Figure 5-5 Automatic Pyrometer

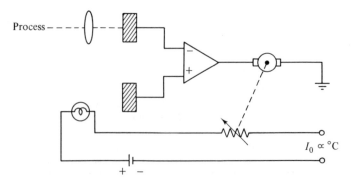

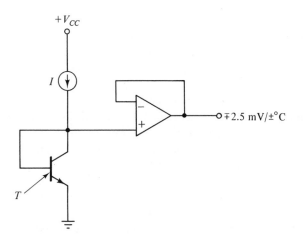

Figure 5-6 Semiconductor Temperature Measurement

5-3 PRESSURE AND FLOW TRANSDUCERS

In the beginning there was the *potentiometric device*. Its low cost and high output have kept it popular in simple systems, but its high sensitivity to shock and vibration and mechanical nonlinearity errors combine to limit its utility. An early technique for overcoming the limitations of potentiometric pressure sensors centered around the *unbonded strain gage*. This device provides substantial improvements in accuracy and stability with typical errors $\frac{1}{2}$% of full scale. However, it is delicate and difficult to fabricate, and its output in the millivolt range usually requires a preamplifier.

An alternative to the unbonded-gage transducer is the *semiconductor gage* bonded directly to the pressure diaphragm, eliminating the mechanical linkages. Frequency response and sensitivity to vibration are improved, and accuracy is equivalent to that of the unbonded implementation. However, its low output also requires a preamplifier, and its low sensitivity makes it suitable only for pressures of 100 psi and above. An improvement is the use of a crystal diaphragm with diffused piezoresistors. The advantage of this technique is its freedom from the measurement hysteresis exhibited by the other methods, since when errors are reduced to the order of $\frac{1}{2}$%, hysteresis becomes the limiting factor.

Present developments in pressure transducers include the incorporation of piezoresistors with hybrid integrated-circuit techniques to compensate for the various error sources. National Semiconductor pioneered this development with their LX-1700 and LX-3700 Series devices, providing an order-of-magnitude price reduction for devices having 1% error. The hybrid device contains a built-in vacuum reference, internal chip heating to minimize temperature effects, and piezoresistors arranged in a Wheatstone bridge sensing circuit with preamplification and signal conditioning included, as shown in Figure 5-7.

Fluid-flow measurement is generally implemented by one of two methods: differential-pressure sensing, and mechanical-contact sensing such as with turbines. Flow rate F is the time rate of fluid motion with typical dimensions in feet per second. Volumetric flow Q is the fluid volume per unit time such as gallons per minute. Mass-flow rate M for a gas is defined in terms of, for example, pounds per hour.

Differential-pressure flow sensing elements have been referred to as *head meters* or *variable-head meters*, because the differential pressure across two measurement points is equated to the head. This is equivalent to the height of the column of a differential manometer. Flow rate can be obtained from its relationship with the 32 ft/sec² gravitational constant g and differential pressure. Liquid flow in open channels is normally obtained by head-producing devices such as flumes and weirs. Head measurement is obtained by measuring the height of the flow over the weir. Volumetric flow can then be obtained by including the cross-sectional area of the flow, as described in Figure 5-8. Figure 5-9 shows examples of differential-pressure sensing elements.

$$\text{flow rate } F \quad = \quad \sqrt{2g} \cdot \Delta P \quad \text{feet/second} \qquad (5\text{-}1)$$

$$\text{volumetric flow } Q \quad = \quad \sqrt{2g} \cdot L \cdot H\sqrt{H} \quad \text{cubic feet/second} \qquad (5\text{-}2)$$

Figure 5-7 Integrated-Circuit Pressure Transducer (*Courtesy National Semiconductor*)

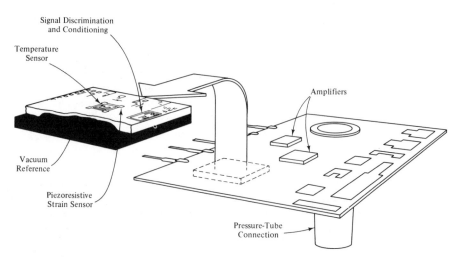

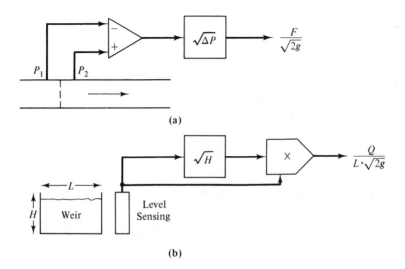

(a)

(b)

Figure 5-8 (a) Flow Rate, (b) Volumetric Flow

The majority of flow-rate, or velocity, measuring instruments are point sensors, such as the pitot tube for gas streams. Single-point measurements, however, are generally inaccurate representations of a process stream. Multipoint-sampling or line-averaging sensors such as an annubar are more accurate. Differential-pressure measurements of gas flows at small pressures are another possible source of uncertainty and may require the application of sensitive LVDT-coupled pressure sensors.

Supplementary static temperature and pressure sensing are required for mass flow-rate measurements. For accuracy, temperature must be measured at each point where velocity is acquired, whereas static pressure can be considered constant across a duct (Figure 5-10). Ideally, velocity sensors should incorporate integral temperature sensors. It is common practice to associate a calibration factor with the probe, as described in the following equation:

$$k = \sqrt{R \cdot \frac{\Delta P_0}{\Delta px}} \quad (°\text{K}/\text{second}^2)^{1/2} \qquad (5\text{-}3)$$

Figure 5-9 Differential Pressure-sensing Device

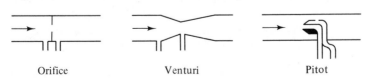

Orifice   Venturi   Pitot

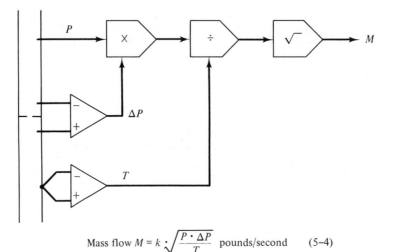

$$\text{Mass flow } M = k \cdot \sqrt{\frac{P \cdot \Delta P}{T}} \quad \text{pounds/second} \quad (5\text{-}4)$$

Figure 5-10 Mass Gas-Flow Computation

where $\quad \Delta P_0 =$ true differential pressure $p_0 - p_\infty$, torr
$\Delta px =$ measured differential pressure
$R =$ universal gas constant

Mechanical-contact flow sensors include turbine and gyroscope transducers, which derive flow rate from angular momentum; thermoelectric
cooling-rate transducers; electrical-resistivity measurements; or beta-decay
methods. The turbine flowmeter, shown in Figure 5-11, is a popular device
that implements equation (5-5) to acquire flow rate:

$$F = \frac{\omega r}{\tan \alpha} \quad \text{feet/second} \qquad (5\text{-}5)$$

where $\qquad \omega =$ the angular rotor velocity
$r =$ the average rotor-blade radius
$\alpha =$ the rotor-blade angle

5-4 DISPLACEMENT SENSORS

Accurate sensing of position, shaft angle, and linear displacement is possible
with the *linear variable-displacement transformer (LVDT)*. With this device an ac
excitation introduced through a variable-reluctance coupling circuit is induced
in an output circuit through a movable core, which determines the amount of
displacement. In comparison with strain-gage transducers the LVDT offers

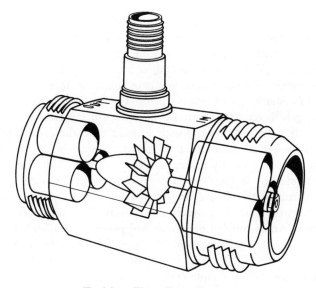

Figure 5-11 Turbine Flow-Rate Transducer

distinct advantages, including overload capability and temperature insensitivity; a disadvantage is its appreciable mass. Sensitivity increases with excitation frequency, but a minimum ratio of 10:1 between excitation and signal

Figure 5-12 Basic LVDT Circuit

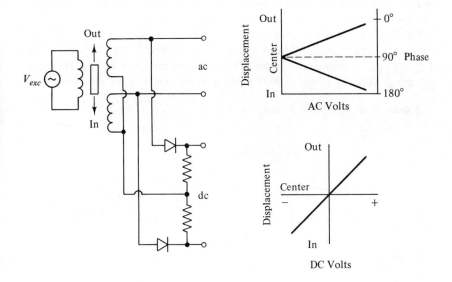

frequencies is considered a practical limit. LVDT variants include the induction potentiometer, synchros, resolvers, and the microsyn. Null balance can be improved with a capacitor across the ac output terminals. Figure 5-12 presents the basic LVDT circuit for both ac and dc outputs.

Acceleration measurements are principally of interest in industrial applications for shock and vibration sensing. Dashpot and capacitive transducers have largely been supplanted by piezoelectric crystals. The equivalent circuit is a voltage source in series with a capacitance, the product of which is a charge in coulombs. As a result of vibratory acceleration an alternating output is generated, typically having a very small amplitude; several crystals are therefore stacked to increase the transducer output. Owing to the small quantities of charge transferred, the transducer is followed by a low-input-bias-current charge amplifier, which also converts the transducer output to a velocity signal. An ac-coupled integrator and precision rectifier will provide a displacement output, which may be calibrated, for example, in millinches of displacement per volt. Figure 5-13 is a block diagram of such a circuit, which is described by the following equations:

$$\text{acceleration} = C \bullet \Delta e \quad \text{coulombs} \tag{5-6}$$

$$\text{velocity } E = \frac{C}{C_f} \bullet \Delta e \quad \text{volts/second} \tag{5-7}$$

$$\text{displacement } V_o = a \int_0^t E \bullet dt \quad \text{volts} \tag{5-8}$$

Hall-effect transducers, which are silicon-substrate based, usually include an integrated amplifier to provide a high output. These devices have a low temperature sensitivity, typically with an operating range of $-40°$ to $+150°$C, and a linear output useful for position sensing and circuit isolation (similar to a photocoupler) such as the Micro Switch LOHET device with a 3.75-mV/gauss response. Figure 5-14 describes the principle of Hall-effect operation. When a magnetic field is set at right angles to an electric field and

Figure 5-13 Vibration Measurement

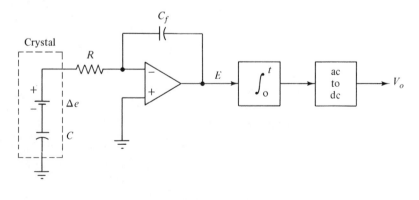

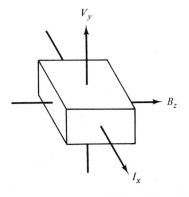

Figure 5-14 Hall-Effect Transducer Principle

both are applied to a current-carrying sample, a force acts on the charge carriers in the sample, creating a diversion of the current flow and a measurable difference of potential across the chip. This V_y voltage is pronounced in materials such as InSb and InAs and occurs to a useful degree in silicon. In applications, the magnetic field is usually provided as a function of the measurand.

Bonded-resistance strain gages are the most widely used transducers because of their versatility, accuracy, and relatively low cost. Initially developed for studies of stress and strain, they have found considerable application in the manufacture of other transducers such as pressure, weight, and vibration sensors. Strain gages may be based on a thin metal wire, foil, thin films, or even a semiconductor. In general, the device includes a force-summing element that changes resistance in proportion to applied pressure. A typical strain element of 350 ohms will register changes to 15 ohms, or 5%. Through a Wheatstone bridge circuit a 10-V reference will be translated into 5-V output with up to a 250-mV change at a stability usually within ±0.1%. The output-terminal impedance is nominally just that of one arm resistance. A bridge's linearity deteriorates with increasing sensitivity, but it can be linearized by the conversion of the four-terminal bridge into a five-terminal circuit and the addition of an operational amplifier. Both of these bridge circuits and their output relationships are shown by Figures 5-15 and 5-16.

Liquid and quasi-liquid (granular solids) levels are process measurements frequently required in tanks, pipes, and other vessels. From level measurement liquid volume and mass can also be determined if the tank geometry and density, respectively, are known. Sensing methods of various complexity are employed in measurements of continuous and incremental-height liquid levels. Float devices and differential-pressure, ultrasonic, and radiation transducers are widely used.

Float devices offer simplicity and various ways of translating motion into

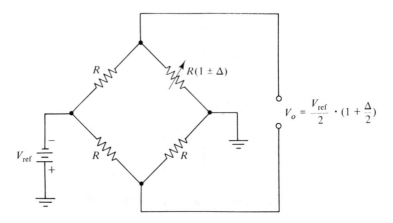

$$V_o = \frac{V_{ref}}{2} \cdot (1 + \frac{\Delta}{2})$$

Figure 5-15 Bridge Transducer Circuit

a level reading. However, they are usually less satisfactory for viscous fluids. A differential-pressure transducer can also measure the height of a liquid when the specific weight of the liquid W is known and the ΔP transducer is connected between the tank surface and bottom. Height is given by the ratio of $\Delta P/W$. Ultrasonic level sensing can be implemented by an echo-ranging system, which is especially useful for tall tanks, and by discrete-height sensing by means of emitters and receivers placed horizontally opposite each other along the tank height. Difficult fluids such as cement and paper-mill digesters are best served by nuclear devices, which operate in much the same way as the discrete-height ultrasonic sensors.

Figure 5-16 Linearized Bridge Circuit

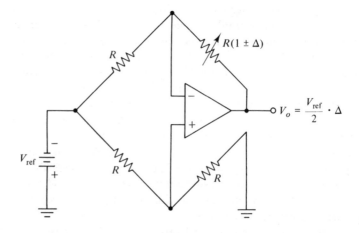

$$V_o = \frac{V_{ref}}{2} \cdot \Delta$$

5-5 *PHOTOMETRY AND RADIOMETRY*

Photometry and radiometry have been described as difficult fields of precise measurement. Much historical confusion has resulted from the use of archaic terms, qualitative definitions, and incorrect substitutions between photometry and radiometry. Planck's basic assumption was that light is not continuous but consists of discrete quanta (photons) whose energy is a function of frequency. For this reason energy E falling on an arbitrary photosensitive material will exhibit a spectral response peak, a function both of the specific material and of equation (5-9). It is therefore necessary to spectrally match sources and sensors to optimize the transfer. This process can be aided by a lens that gathers parallel rays and focuses them on the area of the sensor. The gain associated with this collimation is given by equation (5-10), where the lens transmission loss is approximated as 0.9. Table 5-3 presents common photometric and radiometric definitions, where a source of diameter one-tenth the separation distance is considered an area source. A sphere has a surface area of $4\pi R^2$ and a total solid angle of 4π steradians.

Important sources of radiation are the sun, man-made lamps, lasers, and light-emitting diodes. LED devices can be fabricated to emit on wavelengths between about 560 and 910 nanometers with bandwidths to 30 nanometers. By comparison the response of the human eye peaks at about 555 nanometers. LED's fall into two general classifications, emitters and laser diodes, both of which are photodiodes emitting from their valence bands. The principal difference is that laser diodes have higher peak powers and narrower spectral widths than emitters. Emitters are also usually operated continuously, whereas laser diodes operate only in the pulsed mode. The laser is the only source of coherent radiation and is capable of extremely high radiance. Commonly applied types are summarized in Table 5-4. Plastics are replacing glass for optical transmission. However, both glass and plastic are opaque to ultraviolet below about 350 nanometers, requiring fused quartz or sapphire for low-loss transmission.

TABLE 5-3
PHOTOMETRIC AND RADIOMETRIC DEFINITIONS

PARAMETER	*RADIOMETRY*	*PHOTOMETRY*
Intensity (point source)	Watts/steradian	Candelas
Radiance (area source)	$\dfrac{\text{Watts/steradian}}{\text{cm}^2}$	Footlamberts
Total flux	Watts	Lumens
Irradiance	Watts/cm^2	Footcandles

TABLE 5-4
LASER SUMMARY

TYPE	WAVELENGTH (μm)	PEAK POWER (W)	FEATURE
Argon	0.49	5/100	Blue-green
He-Ne	0.63	0.1/2	Low cost red
CO_2	10.6	200/75 Kw	High power
HeCd	0.44	0.1/2	Recent blue
Krypton	0.64	5/100	Red or green

$$E = hf = (6.626 \times 10^{-34} \text{ joules/second})(f \text{ Hz}) \qquad (5\text{-}9)$$

$$\text{sensor gain} = 0.9 \frac{\text{lens radius}}{\text{sensor radius}} \qquad (5\text{-}10)$$

Light sensors fall into three categories—photoemitters, photodiodes, and photoconductors. Photomultiplier tubes have usable sensitivities down to one photon. Tube gain is expressed by δ^n, which is the ratio of secondary to primary electrons for each of the n dynodes in the tube. Overall gains to 10^6 are common. Photomultipliers are the most sensitive of detectors and require interfacing with low-input-bias-current amplifiers, such as varactor-input-stage operational amplifiers, if this sensitivity is to be fully realized. Equation (5-11) expresses photomultiplier output current as a function of tube gain δ^n:

$$I_O = (1.6 \times 10^{-19}) \cdot \delta^n \quad \text{amperes} \qquad (5\text{-}11)$$

Phototubes, which are in the same class as photomultipliers, were the first devices to operate by cathode-to-anode electron emission upon exposure to incident light. As indicated by Figure 5-17, phototubes are essentially

Figure 5-17 Phototube Characteristics

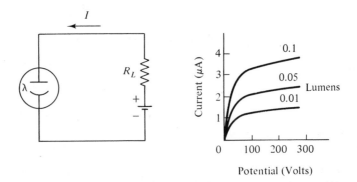

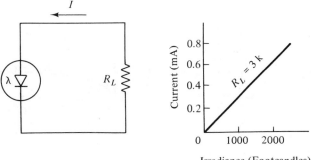

Figure 5-18 Photovoltaic Characteristics

constant-current devices, and a good signal-to-noise ratio is their primary feature. Photovoltaic devices, such as solar cells, provide an output emf of 0.5 volt for silicon and 0.1 volt for germanium with efficiency in the 15% range (Figure 5-18). Maximum power output is obtained by optimizing load resistance, which typically is about 3 k for silicon devices. Operation depends on lowering of the semiconductor potential barrier by incident light, resulting in majority-carrier current flow.

Photodiodes are light-sensitive devices which may either be pn junction diodes or pnp and npn phototransistors; they are among the most widely applied electro-optical sensors. Phototransistors in the Darlington connection are also available for increased sensitivities. However, photodiodes are more linear than phototransistors. Speed, power, and gain tradeoffs between LED sources and photodiode and phototransistor sensors are tabulated in Table 5-5. Signal-transmission applications generally use photodiodes, whereas position sensors and optical isolators typically employ phototransistors. Photoconductive cells are photoresistive devices that exhibit a decreasing resistance with increasing light level. They exhibit hysteresis effects and a temperature coefficient that is a function of light level. Power-dissipation capability must also be observed when applying these devices, which are described by Figures 5-19 and 5-20.

TABLE 5-5
PHOTOSENSOR CHARACTERISTICS

DEVICE	SPEED	OUTPUT	GAIN
Darlington	1 kHz	1 W	1000
Transistor	100 kHz	100 mW	1
Diode	10 MHz	1 mW	0.001

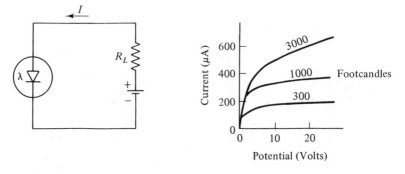

Figure 5-19 Photodiode Characteristics

Figure 5-20 Photoconductive Characteristics

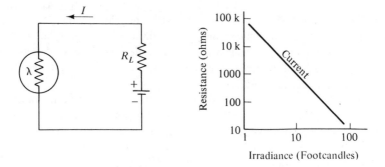

5-6 CHEMICAL ANALYZERS

Online measurements of industrial processes and chemical streams frequently require the use of dedicated analyzers to obtain variables important to the control of a process. Examples are oxygen for boiler control, sulfur oxide emissions from combustion processes, and hydrocarbons found in petroleum refining. Laboratory instruments such as gas chromatographs generally are not used for online measurements, primarily because they analyze all compounds present simultaneously, rather than a single one of interest.

The *dispersive infrared analyzer* is perhaps the most widely used analyzer, owing to the range of compounds it can be configured to measure. Operation is by the differential absorption of infrared energy of the sample stream in comparison to that of a reference cell. Detection is by deflection of a diaphragm separating the sample and reference cells, which, in turn, detunes an oscillator circuit capacitively. *Oxygen analyzers* are usually of the amperometric type in which oxygen is reduced at a gold cathode, resulting in a current flow from a silver anode as a function of this reduction. In a paramagnetic wind device, a wind effect is generated when a mixture containing oxygen produces a gradient in a magnetic field. Measurement is derived by the thermal cooling effect on a heated resistance element, which is a thermal anemometer mechanization.

Hydrocarbon analyzers usually employ the flame ionization method whereby a regulated sample gas passes through a flame fed by regulated fuel and air. Hydrocarbon compounds are thereby ionized into ions and electrons, which are collected and measured with polarized electrodes. *Chemiluminescent-reaction analyzers* produce specific light emission when electronically excited and oxygenated molecules revert to their ground state. These methods are summarized in Table 5-6; a typical analyzer implementation is shown by Figure 5-21.

Electrochemical analytical sensors detect the electrical potential generated in response to the presence of dissolved ionized solids in a process stream. Included in this group are pH, conductivity, and ion-selective probes. The principle of operation is based on the Nernst equation, which typically provides a 60-mV potential change for each tenfold change in the activity of a monovalent ion. For pH sensors, the ion-selective electrode is sensitive to free hydrogen ions in the stream, thereby reflecting the acidity or alkalinity of the sample. pH and oxidation-reduction (ORP) sensors, the latter detecting the ratio of reducing agent to oxidizing agent, are important in effluent measurement and treatment control. Equation (5-12) describes the Nernst relationship:

$$V_o = V + \frac{F}{n} \log (ac + s_1 a_1 c_1 + \cdots) \quad \text{volts} \qquad (5\text{-}12)$$

where
V_o = voltage between sensing and reference electrodes
V = electrode base potential
F = Nernst factor, 60 mV at 25°C
n = ionic charge, 1 monovalent, 2 bivalent, etc.
a = ionic activity
c = concentration
s = electrode sensitivity to interfering ions

TABLE 5-6
CHEMICAL ANALYZER METHODS

COMPOUND	ANALYZER
CO, SO_x, NH_x	Infrared
O_2	Amperometric, paramagnetic
HC	Flame ionization
NO_x	Chemiluminescent
H_2S	Electrochemical cell

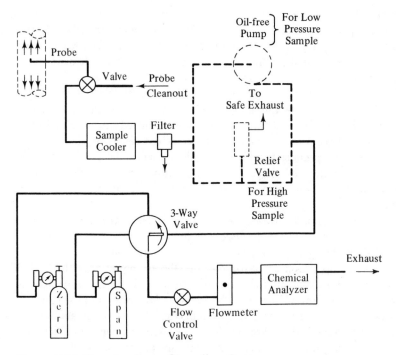

Figure 5-21 Gas-Analyzer Sampling System

pH defines the balance between the hydrogen ions H^+ of an acid and the hydroxyl ions OH^- of an alkali, where one type of ion can be increased only at the expense of another. pH is measured by determining the electrical potential generated in a solution containing these ions with electrodes of special composition. These electrodes, forming the pH probe assembly, are sensitive only to the presence of H^+ ions in the solution, and current flow is kept very low in order to prevent electrode polarization.

REFERENCES

1. W. D. COOPER, *Electronic Instrumentation and Measurement Techniques,* Prentice-Hall, Englewood Cliffs, N.J., 1970.

2. K. ARTHUR, *Transducer Measurements,* Tektronix, Inc., Beaverton, Ore., 1970.

3. D. M. CONSIDINE, *Handbook of Applied Instrumentation,* McGraw-Hill, New York, 1964.

4. *Electrometer Measurements,* Keithley Instruments, 28775 Aurora Road, Cleveland, Ohio, 1972.

5. A. F. GILES, *Electronic Sensing Devices,* Clowes, London, 1966.

6. H. N. NORTON, *Handbook of Transducers for Electronic Measuring Systems,* Prentice-Hall, Englewood Cliffs, N.J., 1969.

7. *Pressure Transducer Handbook,* Bell & Howell, 360 Sierra Madre Villa, Pasadena, Calif., 1974.

8. H. SORENSEN, "Designer's Guide to Optioisolators," *Electronic Design News,* April 5, 1976.

9. L. K. SPINK, *Principles and Practice of Flow Meter Engineering,* Plimpton Press, Norwood, Mass., 1967.

10. *Transducers: Pressure and Temperature,* National Semiconductor, 2900 Semiconductor Drive, Santa Clara, Calif., 1974.

signal acquisition and conditioning

6-0 INTRODUCTION

Analog input systems are assemblages of signal conditioning and conversion devices whose selection is usually based on both performance and economic considerations. They may be functionally partitioned into three subsystems: the input and signal-conditioning circuits; the devices that provide the data-conversion functions; and the digital components that perform the computer interfacing tasks. This chapter provides a detailed development of the first of these subsystems, including a unified method for upgrading signal quality as it progresses from a transducer to the data-conversion system.

A renewed emphasis on data acquisition has resulted from the proliferation of microcomputers and the analog input/output requirements associated with their application to control systems. However, economic considerations are also imposing increased accountability on the design of analog input systems to provide signal-measurement performance only to the required accuracy. Low-level signal conditioning, the more difficult of these design tasks because of the custom nature of acquiring, upgrading, and transmitting analog signals, is dealt with in some detail in this chapter.

The design of data-acquisition systems is largely immune to standardized methods and the attendant cost savings. Nevertheless, a criterion is presented for the organization of these systems based upon performance requirements, and includes transducer-loop interfacing and the tailoring of the signal-conditioning design to dc, sinusoidal, and complex harmonic signal types.

6-1 INPUT GROUNDING, SHIELDING AND TERMINATION

External interference entering low-level instrumentation circuits frequently is substantial, especially in industrial environments, and methods for its prevention or elimination are essential. Noise coupled to transducer leads and input power buses, the primary source of external interference, results from electric- and magnetic-field components radiated by interference sources. For example, unshielded signal cables will couple 1 mV of interference per kilowatt of 60-Hz load for each lineal foot of 1-foot-spaced cable run. Metallic shielding is normally employed to eliminate pickup of these interference components, and for interference below 1 MHz a single-point shield ground is required to prevent circulating currents induced by magnetic coupling effects.

Almost all interference results from near-field sources, defined as being at distances less than one-sixth wavelength of the interfering frequency. In the near field high-voltage low-current sources essentially produce an electric field, and low-voltage high-current sources a magnetic field. For electric fields, which are predominant, the primary attenuation mechanism is reflection by a nonmagnetic material such as copper or aluminum shielding. For magnetic fields absorption is the primary shielding mechanism, and steel or mumetal alloy is required. Magnetic fields are much more difficult to shield against than electric fields, and shielding effectiveness for a given thickness diminishes with decreasing frequency. For example, for steel at 60 Hz realizable magnetic interference attenuation is of the order of 30 voltage dB per 100 mils of shield thickness. Consequently, applications requiring magnetic shielding are normally implemented by cable installation in rigid steel conduit of the necessary thickness. Additional magnetic-field cancellation can be achieved with a twisted cable pair as a result of the periodic pair transposition, provided that the signal-return current is on one of the pair and not on the shield.

For the more common application of shielding against electric-field interference, both copper-foil and braided-shield signal cables offer attenuation of the order of 90 voltage dB to 60-Hz interference. This attenuation decreases by 20 dB per decade of increasing frequency, with the lower resistance of the braided shield being more effective at higher interference frequencies. Crosstalk between adjacent individually shielded low-level instrumentation signal cables is typically of the order of -120 dB. Also, the cable shield and transducer common lead are normally grounded at the transducer at a single point where interference is greatest and provision of the lowest-impedance ground point is most effective. For applications where the transducer is floating or the shield is driven by an amplifier guard drive signal, the ground connection shown in Figure 6-1 is omitted. However, floating-instrumentation amplifiers require a dc-restoration path to ground for input bias currents provided by balanced resistors to ground, whereas isolation amplifiers do not, because they furnish their own bias current to the transducer loop. Table 6-1 summarizes interference-shielding criteria.

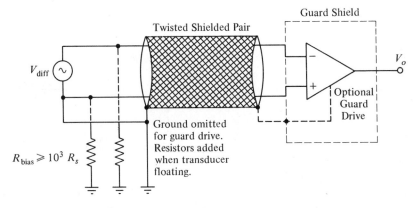

Figure 6-1 Transducer-Loop Termination

Many transducers require excitation in order to convert a measurand into an output signal. Some require ac excitation such as an LVDT for displacement sensing, others dc excitation such as bridges and RTD devices for temperature sensing. DC excitation requirements may be either constant voltage for bridge-circuit transducers, or constant current for RTD temperature transducers. For example, a Sostman four-wire 100-ohm platinum RTD transducer (PT139AX) provides precision temperature measurement over a $-100°$ to $+600°C$ range with a constant 1-mA dc excitation. The four-wire configuration shown in Figure 6-2 eliminates measurement errors resulting from voltage drop in the lead wires. Excitation potentials are usually kept as low as possible to avoid transducer self-heating error effects. A six-wire bridge transducer circuit used for remote strain-gage/pressure sensing, such as the BLH Electronics DHF Series device, allows remote sensing to provide voltage regulation of the excitation power supply at the transducer. The 20-k balance potentiometer permits nulling of any residual offset that may exist. Precision regulation and a very low temperature coefficient are basic requirements for transducer power supplies. Performance to 0.01% is typical and is available with devices such as the Analog Devices Model 2B35 and National Semiconductor LH0070 devices. AC sinewave transducer excitation oscillators with

TABLE 6-1
INTERFERENCE-SHIELDING CRITERIA

SOURCE	INTERFERENCE	SHIELD	MATERIAL	ATTENUATION (60 Hz)
High-V, low-I	Electric field	Reflective	Copper or aluminum	90 dB
High-I, low-V	Magnetic field	Absorptive	Steel or mumetal	30 db/ 100 mils

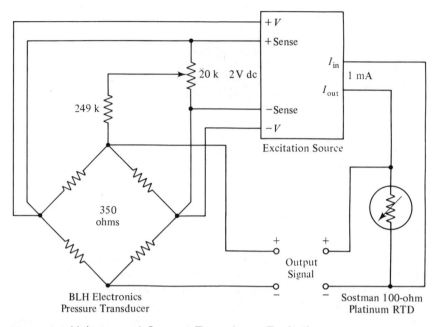

Figure 6-2 Voltage and Current Transducer Excitation

frequency and amplitude stability of the same order are also available, as represented by the Burr Brown 4023 oscillator.

Data-acquisition systems are usually interfaced to the physical environment with a connection panel, such as a screw-terminal barrier strip, and a connection matrix that will accommodate specific user input options on a per-channel basis. These options can include 250-ohm current loop resistors for converting single-ended 4-to-20mA signals to $1-5$ V, input filters, transducer excitation and open-detection circuits, voltage dividers, thermocouple cold-junction compensation, amplifier input-bias-current return resistors for floating transducers, and protection circuits. Figure 6-3 illustrates this interface for a differential transducer showing a 1-Hz cutoff one-pole RC lowpass differential filter, described later, and an open-transducer detection circuit that results in a negative full-scale input for a broken signal lead. During normal operation the $-V$ is dropped across the 100-M resistor and results in a negligible differential voltage error of only $-V \cdot R_s/100$ M. However, it should be obvious that use of this circuit prevents the realization of CMRR because of the severe differential imbalance produced by the 10-k and 100-M resistors. Optional input protection circuits shown include fuses to clear catastrophic input faults for hazardous applications, and zener transient suppressors to clip fast-risetime overvoltages. Varistors such as MOV devices are not suitable for this application because of their leakage current. Cold-

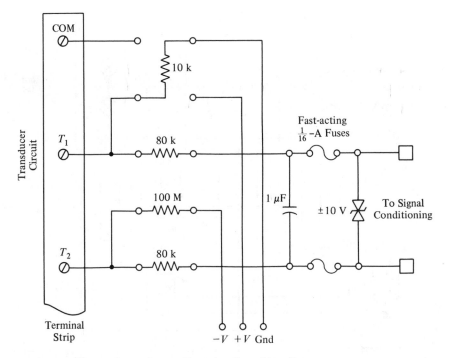

Figure 6-3 Transducer Input Termination Circuit

junction thermocouple compensation circuits such as shown in Figure 5-3 may also be included in this interface.

Mutual coupling between circuits of a data-acquisition system resulting from finite signal-path and power-supply impedances are additional sources of interference. An effective solution is to separate analog signal grounds from noisier digital and chassis grounds by separate ground returns, all terminated at a common single-point ground, as illustrated in Figure 6-4. Soldered or brazed ground connections are always preferred to bolted ones to prevent high-resistance connections, and caution should be exercised when joining dissimilar metals because of the possibility of galvanic corrosion.

6-2 SIGNAL-CONDITIONING ERROR ANALYSIS

Since a transducer output signal corresponds to some measurand of interest, such as temperature or pressure, of primary concern is the acquisition of a true amplitude measurement a. A sinusoidal phase measurement is also

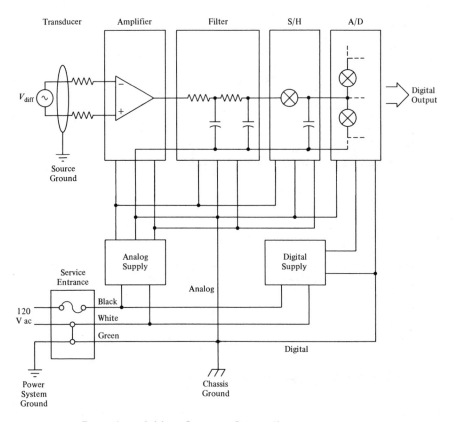

Figure 6-4 Data-Acquisition-System Grounding

occasionally required, such as from a resolver or angle encoder, where measurement of the true phase angle ϕ is of interest. Closed-form expressions are available for determining the accuracy of an amplitude or phase measurement, which really are estimates when the signals are corrupted by interference, but require the assumption of matched-filter processing. This theoretical embodiment maximizes the signal-to-noise ratio (SNR) by tailoring its frequency-response function to a specific signal, thereby passing the signal and excluding noise. SNR is a dimensionless ratio of watts of signal to noise, provides a useful criterion of merit of an information system for performance-evaluation purposes, and will be adopted in this text.

The probability, or confidence, that a signal corrupted by interference is within a specified Δ region centered about the true value for either amplitude or phase may be represented by equations (6-1) and (6-2). These equations are exact for sinusoidal signals in random Gaussian noise and are asymptotically correct, depending upon how well the signal-plus-noise conforms to the exact case. Nevertheless, these results are conservative, since most signals can be decomposed into sinusoidal components, and the central-limit theorem provides that the summation of a number of independent noise sources tends to a Gaussian probability density in the limit. Table 6-2 presents a tabulation from substitution into these equations for amplitude and phase errors for a 68% (1σ) confidence in their measurement at specified SNR values, where the Δ ratios correspond to the error values. One sigma is a reasonable confidence level for many applications. For 95% (2σ) confidence the error values for each entry are doubled for the same SNR.

$$P(\Delta a; a) = \text{erf}\left(\frac{1}{2} \cdot \frac{\Delta a}{a} \cdot \sqrt{SNR}\right) \qquad (6\text{-}1)$$

$$P(\Delta\phi; \phi) = \text{erf}\left(\frac{1}{2} \cdot \frac{\Delta\phi}{57.3^0/\text{rad}} \cdot \sqrt{SNR}\right) \qquad (6\text{-}2)$$

The foregoing SNR requirements related to amplitude and phase errors in matched-filter signal processing are of considerable utility for determining the accuracy of measurement signals. To be of practical use, however, they must be mathematically related to amplifier and linear filter signal-conditioning circuits. Figure 6-5 describes the basic signal-conditioning structure, *linear* including the important preconditioning amplifier and postconditioning filter *vs* bandwidths and amplifier CMRR. Earlier work by Fano[1] shows that under *matched* high-input SNR conditions linear filtering approaches matched filtering in its *filters*

? ?

TABLE 6-2
SNR VERSUS AMPLITUDE AND PHASE ERRORS *(for 1σ)*

SNR	AMPLITUDE ERROR (%FS)	PHASE ERROR (DEG)	CONFIDENCE (%)
10^1	44.0	22.5	68
10^2	14.0	7.5	68
10^3	4.4	2.5	68
10^4	1.4	0.82	68
10^5	0.44	0.27	68
10^6	0.14	0.09	68
10^7	0.044	0.027	68
10^8	0.014	0.009	68
10^9	0.0044	0.003	68
10^{10}	0.0014	0.001	68
10^{11}	0.00044	0.0003	68
10^{12}	0.00014	0.0001	68

Double error values for 2σ; 95% confidence

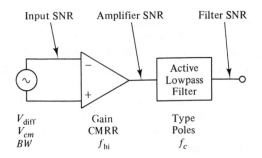

Figure 6-5 Signal-Conditioning Structure

efficiency. Later work by Budai[2] developed a relationship for this efficiency, expressed by the characteristic curve of Figure 6-6. This K curve appears most reliable for amplifier numerical output SNR values between about 10 and 100, with a K of 0.9 for SNR values of 100 and greater a practical upper limit.

Equations (6-3) through (6-6) describe the relationships upon which the improvement in signal quality by linear filtering is based for both random and coherent interference. Both rms and dc voltage values may be used in equation (6-3). CMRR is squared in equation (6-4) in order to convert its ratio of voltage gains dimensionally to the power ratio represented by SNR. Equation (6-5) represents the processing-gain relationship for random interference. In general, the reason why the increasing amplifier bandwidth of this equation

Figure 6-6 Linear Filter Efficiency K Versus SNR

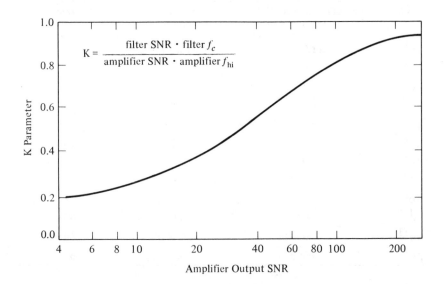

increases the filter output SNR is the decorrelation effect on the instantaneous vector relationships of signal-plus-random-noise as the bandwidth they simultaneously occupy is increased.[3] The processing gain will be effective in further improving signal quality over that provided by the amplifier common-mode rejection ratio if the interference is largely random in nature. The majority of improvement is provided by the amplifier CMRR, which is especially capable of attenuating lower-frequency coherent interference such as at 60 Hz. The residual amplifier noise output usually consists of higher-frequency random components, against which CMRR has limited effectiveness, but linear filtering is capable of providing further SNR improvement.

Under coherent interference conditions additional signal-quality improvement is a function of achievable filter attenuation in relation to the interference frequency expressed by equation (6-6). For example, 60-Hz interference and a 100-Hz filter-passband requirement preclude signal-quality improvement beyond that provided by the amplifier CMRR. In contrast, a 1-Hz filter-passband requirement for a signal permits substantial attenuation of the 60-Hz interference from filtering, but this direct attenuation mechanism has no relationship to the processing gain of equation (6-5). Note that the filter cutoff frequency is determined by signal requirements presented in Tables 1-2 and 4-9. An additional consideration that must be evaluated is the contribution to the overall error budget by the filter component error from Table 4-9, which is developed in Section 4-4.

An easily confused relationship is that between the error budget of a signal-conditioning channel and its output-signal quality. The amplitude or phase quality of the output signal is determined by elements that upgrade this quality, such as filtering, and these are independent of component errors such as those contributed by the transducer, amplifier, and filter. Signal quality that is adequate for the measurement accuracy of interest is a necessary but not sufficient condition. Therefore, the rms combination of both component and signal-quality errors determines the overall measurement accuracy, where the former is usually dominant in this outcome.

Examples presented in the sections that follow illustrate these relationships, using the amplifier component errors ϵ_{ampl} from Table 3-5 and filter component errors ϵ_{filter} from Table 4-9. Equations (6-3) through (6-6) with Table 6-2 permit determination of the available signal-quality improvement terms, which are then combined with the data-acquisition channel component errors by equations (6-7) or (6-8). The $\epsilon_{SNR_{ampl}}$ derived from equation (6-4) and Table 6-2 is equal to the amplifier common-mode error term $\epsilon \sqrt{2}\, V_{cm}/$ CMRR contained in ϵ_{ampl} from Table 3-5. Consequently, with the application of linear filtering $\epsilon_{SNR_{ampl}}$ is subtracted from ϵ_{ampl} and then amended by the upgraded terms $\epsilon_{SNR_{random}}$ and $\epsilon_{SNR_{coherent}}$ if both sources of interference are present and improvement is realized. Applications in which a filter is necessary, such as for data-conversion antialiasing purposes, but no additional SNR upgrading is realized over that provided by the amplifier, merely require equation (6-8) to be employed with deletion of the third, fifth, and sixth terms.

$$\text{input SNR} = \left(\frac{V_{\text{diff}}}{V_{cm}} \right)^2 \qquad \text{(6-3)}$$

DC and RMS applicable

$$\text{amplifier SNR} = \text{input SNR} \cdot \text{CMRR}^2 \qquad \text{(6-4)}$$

$$\text{filter SNR}_{\text{random}} = \text{amplifier SNR} \cdot K \cdot \frac{\text{amplifier } f_{\text{hi}}}{\text{filter } f_c} \qquad \text{(6-5)}$$

$$\text{filter SNR}_{\text{coherent}} = \frac{\text{amplifier SNR}}{(\text{filter attn.})^2} \qquad \text{(6-6)}$$

$$\epsilon_{\text{channel}\%\text{FS}}\big|_{\text{no filter}} = [\epsilon_{\text{sensor}}^2 + \epsilon_{\text{ampl}}^2]^{1/2} \qquad \text{(6-7)}$$

$$\epsilon_{\text{channel}\%\text{FS}}\big|_{\text{filter}} = [\epsilon_{\text{sensor}}^2 + \epsilon_{\text{ampl}}^2 - \epsilon_{\text{SNR}_{\text{ampl}}}^2 \qquad \text{(6-8)}$$
$$+ \epsilon_{\text{filter}}^2 + \epsilon_{\text{SNR}_{\text{random}}}^2 + \epsilon_{\text{SNR}_{\text{coherent}}}^2]^{1/2}$$

6-3 DC AND SINUSOIDAL SIGNAL-CONDITIONING EXAMPLES

An initial step is to evaluate the characteristics of the input signal and noise to determine their type and the input SNR. The signal type will determine the choice of linear filter if this component is required in the signal-conditioning channel. The dc or rms transducer signal voltage is normally available from the manufacturer for the range of intended operation. Note that the maximum expected transducer voltage is employed because of the necessity to relate this to the permissible full-scale output level from the channel. After calculation and design of the signal-conditioning channel, the signal-quality and component errors combined by equation (6-8) define the minimum usable signal amplitude as a percentage of the full-scale output level. This term was defined by equation (3-19) for the amplifier alone. The interference voltage is measured between the signal leads and ground with the transducer disconnected and the signal cable terminated in the transducer source impedance. A true rms responding voltmeter is the best choice for obtaining this measure-ment because of the relatively high crest factor (the ratio of the peak to rms value) of some forms of interference. However, an average responding ac voltmeter calibrated for sine-wave rms values, the type typically available, may be used, but its reading must be multiplied by 1.127 for random interference. The signal used in the following examples is dc, but the design procedures are the same for sinusoidal signals.

Consider a thermocouple signal-conditioning channel subjected to operating conditions corresponding to those for the instrumentation amplifiers in Table 3-5 including an $A_{v_{diff}}$ of 10^3 and combined random plus 60-Hz V_{cm} interference of 1 V_{rms}. The greatest improvement in measurement accuracy is available from the choice of amplifiers that have low systematic and random errors in comparison with their common-mode error. An example is the single-instrumentation-amplifier AD542 device (Figure 6-7). The type-S thermocouple transducer is useful for high-temperature applications to 1000°C, where it has a 10-mV dc output signal at 0.25% error, as described in Table 5-2. However, the application of post-signal-conditioning signal processing can linearize this thermocouple to an accuracy of 0.1% if it is required. This result is usually achieved by a table look-up digital linearization routine resident in an associated microcomputer.

Substitution of these signal and common-mode interference voltages into equations (6-3) and (6-4) for the AD542 amplifier, possessing an incircuit CMRR of 0.833×10^4 from Figure 3-15, provides an amplifier SNR of 7×10^3. Upon interpolation of Table 6-2, this $\epsilon_{SNR_{ampl}}$ term exhibits a 1-σ amplitude error of 1.7%. Note that this error is identical to the 170 μV (1.7% of FS) shown on line 5 of Table 3-5 for this device, as expected. Without linear filtering, however, only the transducer and amplifier component errors are involved in the signal-conditioning error budget. This is represented by equation (6-7) and provides a total output error of 2.0% which includes the ϵ_{ampl} error from Table 3-5. The three-pole Butterworth filter shown in Figure

Figure 6-7 AD542 Gain Versus Bandwidth

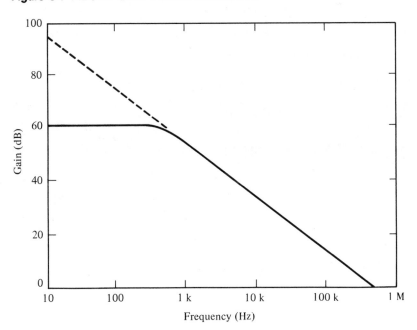

Frequency (Hz)

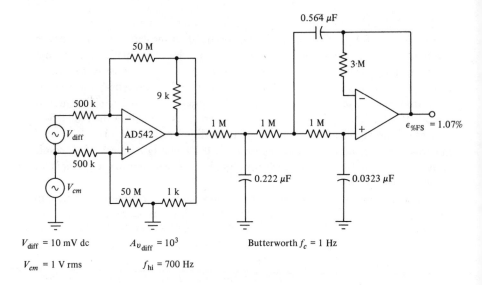

V_{diff} = 10 mV dc $\qquad$ $A_{v_{\text{diff}}} = 10^3$ $\qquad\qquad$ Butterworth f_c = 1 Hz

V_{cm} = 1 V rms $\qquad\qquad$ f_{hi} = 700 Hz

Figure 6-8 Low-Cost Signal-Conditioning Channel

$$\text{input SNR} = \left(\frac{V_{\text{diff}}}{V_{\text{cm}}}\right)^2 \qquad\qquad (6\text{-}3)$$

$$= \left(\frac{10^{-2}\,V}{1\,V}\right)^2 = 10^{-4}$$

V diff = 10 mV ?

$$\text{amplifier SNR} = \text{input SNR} \cdot \text{CMRR}^2 \qquad\qquad (6\text{-}4)$$

$$= (10^{-4}) \cdot (0.833 \times 10^4)^2$$

$$= 7 \times 10^3 \quad (1.7\% \text{ @ } 1\,\sigma)$$

$$\text{filter SNR}_{\text{random}} = \text{amplifier SNR} \cdot \text{K} \cdot \frac{\text{amplifier}\,f_{\text{hi}}}{\text{filter}\,f_c} \qquad\qquad (6\text{-}5)$$

$$= (7 \times 10^3)(0.9)\frac{(700\text{ Hz})}{1\text{ Hz}}$$

$$= 4.9 \times 10^6 \quad (0.075\% \text{ @ } 1\,\sigma)$$

$$\text{filter SNR}_{60\text{ Hz}} = \frac{\text{amplifier SNR}}{(\text{filter attn.})^2} \qquad\qquad (6\text{-}6)$$

$$= \frac{(7 \times 10^3)}{(10^{-5})^2} = 7 \times 10^{13} \quad (0.0001\% \text{ @ } 1\,\sigma)$$

$$\epsilon_{channel\%FS}\big|_{no\ filter} = [\epsilon_{sensor}^2 + \epsilon_{ampl}^2]^{1/2} \qquad (6\text{-}7)$$

$$= [(0.25\%)^2 + (1.98\%)^2]^{1/2}$$

$$= 2.0\%$$

$$\epsilon_{channel\%FS}\big|_{filter} = [\epsilon_{sensor}^2 + \epsilon_{ampl}^2 - \epsilon_{SNR_{ampl}}^2 \qquad (6\text{-}8)$$

$$+ \epsilon_{filter}^2 + \epsilon_{SNR_{random}}^2 + \epsilon_{SNR_{60\ Hz}}^2]^{1/2}$$

$$= [(0.25\%)^2 + (1.98\%)^2 - (1.7\%)^2$$

$$+ (0.2\%)^2 + (0.075\%)^2 + (0.0001\%)^2]^{1/2}$$

$$= 1.07\%$$

6-8 was then selected because it provides the proper characteristic for dc-type signals, and sinusoidal signals also when encountered. Substitution into equation (6-5) for this circuit with a 1-Hz filter cutoff and an efficiency factor K of 0.9, obtained from Figure 6-6 for a 7 × 10^3 amplifier SNR, results in a filter SNR of 4.9 × 10^6. This corresponds to a signal quality of 0.075% amplitude error due to random interference. Considering next the 60-Hz coherent interference, the 1-Hz-cutoff lowpass filter provides an attenuation of approximately 10^{-5} at 60 Hz obtained from the Butterworth amplitude response curves of Chapter 4. Equation (6-6) then provides a filter SNR of 7 × 10^{13} for an error of 0.0001%, but the residual noise of the filter may limit the maximum realizable improvement. The previously determined 1.7% amplifier common-mode error is subtracted from the 1.98% AD542 component error from Table 3-5, and both the improved signal quality and filter component errors are combined by equation (6-8). For this example the realizable channel total output error including the linear filtering improvement is found to be 1.07%. The factor-of-two net increase in measurement accuracy for this low-cost signal-conditioning channel is notable, especially considering the addition of the 0.2% filter component error and a nonlinearized thermocouple. *table 4.9*

Greater measurement accuracy can be achieved, as shown in Figure 6-9, by employing a high-performance ICL7605 CAZ instrumentation amplifier with a fixed −20-dB/decade rolloff 10-Hz bandwidth and 10^5 CMRR. This amplifier provides nearly half an order-of-magnitude improvement in measurement accuracy in comparison with the AD542 device, but at a cost premium. The measurement accuracy of this dc signal-conditioning channel with or without the addition of linear filtering is essentially unchanged for random and coherent interference, where the realizable signal improvement will approximately compensate for the added filter error. The only motivation for including a linear filter, therefore, would be in sampled-data applications requiring a sharper cutoff antialiasing filter than the one-pole response provided by the 10-Hz amplifier bandwidth.

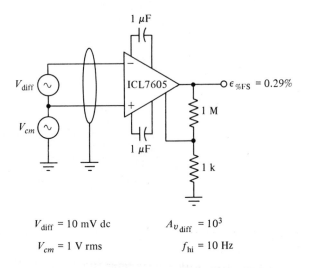

$$V_{diff} = 10 \text{ mV dc} \qquad A_{v_{diff}} = 10^3$$
$$V_{cm} = 1 \text{ V rms} \qquad f_{hi} = 10 \text{ Hz}$$

Figure 6-9 CAZ Signal-Conditioning Channel

$$\text{input SNR} = \left(\frac{10^{-2} \text{ V}}{1 \text{ V}}\right)^2 = 10^{-4} \qquad (6\text{-}3)$$

$$\text{amplifier SNR} = (10^{-4}) \cdot (10^5)^2 = 10^6 \quad (0.14\% \text{ @ } 1 \text{ } \sigma) \qquad (6\text{-}4)$$

$$\epsilon_{channel\%FS}\Big|_{no\ filter} = [\epsilon_{sensor}^2 + \epsilon_{ampl}^2]^{1/2} \qquad (6\text{-}7)$$

$$= [(0.25\%)^2 + (0.15\%)^2]^{1/2}$$

$$= 0.29\%$$

6-4 COMPLEX HARMONIC SIGNAL-CONDITIONING EXAMPLES

Consider a complex harmonic signal with a 100-Hz data bandwidth and again the amplifier errors tabulated in Table 3-5. The previous 10-mV dc signal is reduced to 7-mV rms to prevent exceeding 10 V-peak outputs, which results in only minor error in the use of ϵ_{ampl} values from the table. In accordance with Table 4-9 a Bessel lowpass filter is used with a cutoff frequency of 2.5 times the 100-Hz data bandwidth in order to achieve a 0.75% average filter error. A three-pole design provides an adequate cutoff characteristic, where the V_{cm} is random interference without coherent components. An AD288 transformer-coupled isolation amplifier is utilized with an $A_{v_{diff}}$ of 10^3 at a bandwidth of 200 Hz and incircuit CMRR of 10^5. The calculations provide no improvement in signal quality in this specific example as a result of linear filtering beyond

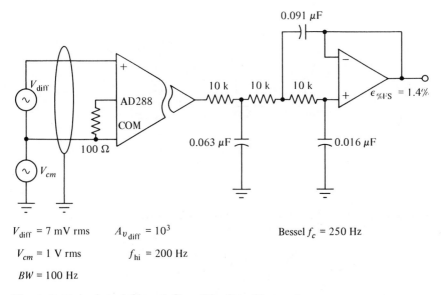

V_{diff} = 7 mV rms $\quad A_{v_{\text{diff}}}$ = 10^3 $\qquad\qquad$ Bessel f_c = 250 Hz

V_{cm} = 1 V rms $\qquad f_{\text{hi}}$ = 200 Hz

BW = 100 Hz

Figure 6-10 Isolated Signal-Conditioning Channel

$$\text{input SNR} = \left(\frac{V_{\text{diff}}}{V_{cm}}\right)^2 = \left(\frac{7 \times 10^{-3}\,\text{V}}{1\,\text{V}}\right)^2 = 4.9 \times 10^{-5} \quad (6\text{-}3)$$

$$\text{amplifier SNR} = \text{input SNR} \cdot \text{CMRR}^2 \qquad\qquad\qquad (6\text{-}4)$$

$$= (4.9 \times 10^{-5})(10^5)^2 = 4.9 \times 10^5 \quad (0.2\% \text{ @ } 1\,\sigma)$$

$$\text{filter SNR}_{\text{random}} = \text{amplifier SNR} \cdot \text{K} \cdot \frac{\text{amplifier}\,f_{\text{hi}}}{\text{filter}\,f_c} \qquad (6\text{-}5)$$

$$= (4.9 \times 10^5)(0.9)\left(\frac{200\,\text{Hz}}{250\,\text{Hz}}\right)$$

$$= 3.5 \times 10^5 \quad (0.2\% \text{ @ } 1\,\sigma)$$

$$\epsilon_{\text{channel}\%\text{FS}}\big|_{\text{filter}} = [\epsilon_{\text{sensor}}^2 + \epsilon_{\text{ampl}}^2 + \epsilon_{\text{filter}}^2]^{1/2} \qquad (6\text{-}8)$$

$$= [(1.0\%)^2 + (0.43\%)^2 + (0.75\%)^2]^{1/2}$$

$$= 1.4\%$$

that provided by the amplifier. With a 1% transducer error the overall signal-conditioning channel error is determined to be 1.4%. This signal-conditioning circuit, shown by Figure 6-10, illustrates data acquisition that also satisfies data-conversion antialiasing requirements.

Recognition of the filter amplitude error in the conditioning of complex harmonic signals has led to the development of minimum-amplitude-error filters, represented by Electronic Instrumentation and Technology's Bessel-

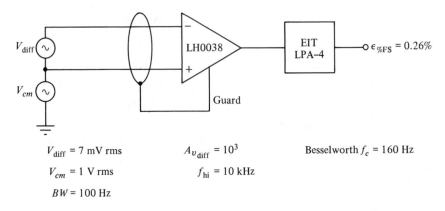

$$V_{diff} = 7 \text{ mV rms} \qquad A_{v_{diff}} = 10^3 \qquad \text{Besselworth } f_c = 160 \text{ Hz}$$

$$V_{cm} = 1 \text{ V rms} \qquad f_{hi} = 10 \text{ kHz}$$

$$BW = 100 \text{ Hz}$$

Figure 6-11 Premium Signal-Conditioning Channel

$$\text{input SNR} = \left(\frac{7 \times 10^{-3} \text{ V}}{1 \text{ V}} \right)^2 = 4.9 \times 10^{-5} \qquad (6\text{-}3)$$

$$\text{amplifier SNR} = (4.9 \times 10^{-5})(10^6)^2 \qquad (6\text{-}4)$$

$$= 4.9 \times 10^7 \quad (0.02\% \text{ @ } 1 \text{ } \sigma)$$

$$\text{filter SNR}_{random} = (4.9 \times 10^7)(0.9) \left(\frac{10 \text{ kHz}}{160 \text{ Hz}} \right) \qquad (6\text{-}5)$$

$$= 27.5 \times 10^8 \quad (0.003\% \text{ @ } 1 \text{ } \sigma)$$

$$\epsilon_{channel\%FS}\big|_{filter} = [\epsilon_{sensor}^2 + \epsilon_{ampl}^2 - \epsilon_{SNR_{ampl}}^2 \qquad (6\text{-}8)$$

$$+ \epsilon_{filter}^2 + \epsilon_{SNR_{random}}^2]^{1/2}$$

$$= [(0.1\%)^2 + (0.13\%)^2 - (0.02\%)^2$$

$$+ (0.2\%)^2 + (0.003\%)^2]^{1/2}$$

$$= 0.26\%$$

worth device described in Chapter 4. Coordination of this four-pole filter with the National Semiconductor LH0038 instrumentation amplifier is shown in Figure 6-11, where the same signal source is considered with an amplifier bandwidth of 10 kHz and CMRR of 10^6 at the $A_{v_{diff}}$ of 10^3 described in Table 3-2. In the following calculations the transducer is assumed linearized to 0.1% in order to demonstrate achievable accuracy in a premium signal-conditioning channel.

From the foregoing examples it is apparent that the greatest improvement in a conditioned signal achievable with linear filtering and the process-

TABLE 6-3
REPRESENTATIVE DATA-ACQUISITION-SYSTEM ERRORS

Signal Type	ERROR (%FS)				
	Transducer	Amplifier	Filter	Signal	RMS Total
DC and	0.1%	0.1%	0.2%	0.01%	0.25%
sinusoidal	1.0%	0.5%	Butterworth	0.1%	1.15%
Complex harmonic	0.1%	0.1%	0.2% Besselworth	0.01%	0.25%
	1.0%	0.5%	0.75% Bessel	0.1%	1.35%

ing-gain relationship is available with instrumentation amplifiers having low offset and random errors in comparison with their CMRR. In other cases the signal-conditioning channel error is neither improved nor degraded by the addition of linear filtering, but the additional signal quality achieved succeeds in compensating for the added filter component error. With an overall channel error budget defined by equation (6-8), a tradeoff analysis is possible among the various elements that contribute to this budget. However, the filter requirement is essential for antialiasing purposes in sampled-data applications regardless of the signal-quality improvement; this topic is developed in Section 8-1. Table 6-3 summarizes expected data-acquisition-system performance under realistic operating conditions and with contemporary devices. These results may be utilized for upgrading analog-controller and chart-recorder sensor measurements or combined with the additional errors associated with data-conversion systems as developed in Chapter 8.

Overload detection is a useful feature where high-gain amplification or the possibility of large-amplitude interfering signals is involved. The LED in the circuit of Figure 6-12 will indicate when the instrumentation amplifier output exceeds $+10.02$ V peak. It does so by sensing the direction of the current flowing between the input network and the amplifier summing junction. The output is approximately logarithmic, providing a graded indication, and other trip points may be provided by altering the reference voltage divider. The two-stage diode-bound circuit prevents overload-recovery delay with the detector amplifier following a state change.

6-5 *DATA-ACQUISITION-SYSTEM ORGANIZATION*

The requirements of the input circuits for a data-conversion system are to receive measurement signals from a data-acquisition system at the accuracy of

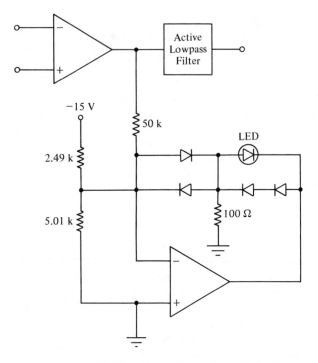

Figure 6-12 +10-Volt Signal Overload Detector

interest, and possibly to provide input isolation to prevent catastrophic voltages from entering the system. Three alternative input configurations including signal conditioning are described by Table 6-4. These configurations are discussed below, and illustrative examples are given by Figures 6-13 through 6-15.

TABLE 6-4
DATA-CONVERSION-SYSTEM INPUT CONFIGURATIONS

| PERFORMANCE | SIGNAL TYPES | SIGNAL CONDITIONING | | |
		Method	CMRR	Filter
Premium	DC, sinusoidal, complex harmonic	Instrumentation amplifier	10^6	Active
Intermediate	DC, sinusoidal	Flying capacitor	10^5	RC
Economy	DC	V/F converter	None	Averager

The premium performance arrangement, which offers the best specifications and flexibility, normally is considered when either low-level or wide-bandwidth signals are involved. The methods of Section 6-2 provide the low-level signal-conditioning accuracies summarized in Table 6-3 for the various signal types indicated. The premultiplexer amplifier-filter circuit per channel represents the optimum input configuration, because the signal conditioning can then be tailored to each transducer. The alternative of sharing a postmultiplexer programmable amplifier-filter among multiple low-level signals presents greater difficulty because of the time required to achieve steady-state following each multiplexer channel transition. And this problem is magnified with frequently encountered low-frequency instrumentation signals.

Common-mode voltages (CMV) result from potential differences between grounded transducers and data-acquisition-system ground references, such as with bridge circuits, and induced interference coupled to signal cables. CMV up to ±10 volts can be accommodated by the rejection capabilities of operational or instrumentation amplifiers when operated from conventional ±15-V power supplies. Beyond this CMV level isolation can be provided, typically to ±1500 V peak, for off-ground signal measurements and protection against high-voltage faults by the isolation amplifier presented in Section 3-1. The front end of the isolation amplifier, normally equivalent in performance to a three-amplifier instrumentation amplifier, is adequate for all but the most difficult signal-conditioning tasks. To provide integrity of the isolation barrier, the amplifier front end must be operated from an isolated dc—dc power supply. In addition, to prevent frequency beats between the internal oscillators of individual dc—dc power supplies, it is common practice to synchronously excite these isolated power supplies from a common driver when multiple isolation amplifiers are employed in a data-acquisition system.

Such an arrangement is described by Figure 6-13, showing synchronized isolation amplifiers used both as an isolated instrumentation amplifier and as a unity-gain isolator, the latter also supplying isolated power for a transducer preamplifier. Note that the isolation-amplifier shield connection terminates in

ISOLATION		MULTIPLEXER
Method	Level	
Isolation amplifier	±1500 V	Analog
Flying capacitor	±250 V	Flying capacitor
Optical coupler	±1500 V	Digital

a floating internal shield, which is grounded at the source with the cable shield. Also, the common-mode signal guard drive of the transducer preamplifier preserves the cable CMRR, as discussed in Section 3-3. The active filter type and parameters appropriate for the specific signal and presampling antialiasing requirements of each input channel are summarized in Tables 4-9 and 8-1. Preamplifier input-bias-current return resistors for floating transducers are included as required.

The premium data-acquisition input circuits of Figure 6-13 suggest the concentration of the electronic devices at a central location, with remote connection to transducers using shielded twisted-pair twinax instrumentation cable. This is a viable arrangement for distances of 100 feet or so, and it has the advantage of meeting the requirements for powering all of the data-acquisition-system devices at a single location. Alternately, instrumentation amplifiers can be located at the transducers and high-level, normal-mode signals sent to the data-acquisition-system termination. However, this does not extend the practical transmission distance beyond about 100 feet, requires provision for remote power at each amplifier, and still necessitates an isolation

Figure 6-13 Premium Data-Acquisition Circuits

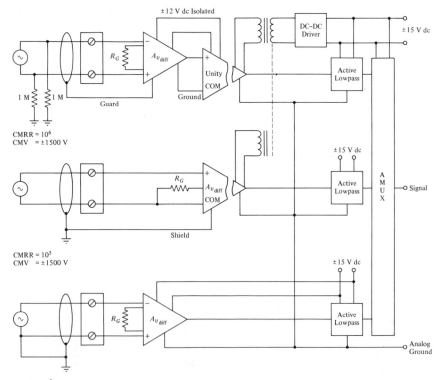

amplifier in each channel if high-CMV protection is necessary. An antialiasing filter is always recommended in sampled-data systems to prevent the conversion of noise at frequencies above the signal and broadening of the sampling spectrum as described in Section 8-1.

The intermediate performance range of input-signal conditioning circuits is represented by the flying-capacitor multiplexer, which can accommodate both high-level and low-level signals. This method is well suited for industrial applications where severe electrical environments demand high noise immunity and the ability to withstand substantial common-mode voltages and where sample-rate requirements are not high for typically encountered dc and sinusoidal signals. In addition to a sample-rate limitation of about 200 Hz due to the dynamic requirements of mechanical switching, reliable operating life is of the order of 10,000 hours at the low sampling rates typically employed with these systems. However, CMRR values to 10^5 are achievable incircuit with CMV protection to ±250 V dc, the latter because of the differential break-before-make switching of flying-capacitor multiplexers. This method is also more cost effective for many input channels in comparison with the per-channel isolation-amplifier active-filter signal-conditioning arrangement but cannot match its flexibility. The flying-capacitor multiplexer is normally connected to the input circuit and tracks the transducer signal with the RC time constant of its circuit, which is essentially that of the RC antialiasing lowpass filter. Consequently, an appropriate number of time constants must be provided after a switching cycle to allow the capacitor to settle to within an acceptable percent error of the signal amplitude. For a ±full-scale signal rollover between samples, 9 RC time constants would be required for settling to 0.012%. However, signal change is usually modest between samples requiring a proportionally shorter time, a topic explored in the low-data-rate analog input system example of Section 8-3.

The simplicity of the passive flying-capacitor data-acquisition input system is apparent from Figure 6-14. Amplification of low-level transducer signals is normally achieved with a postmultiplexer amplifier. For varying signal levels from channel to channel, this amplifier can be a programmable gain device synchronized with the multiplexer address. The flying capacitor serves also as a sample-hold device with an effective aperture time equivalent to relay bounce time, typically in the 1-ms range for reed relays. Thermal-offset-voltage compensation is usually provided for the switching contacts by means of a parallel B_eO internal junction, and capacitor values are typically $1-10 \ \mu F$ as a compromise between acquisition-mode RC charging and hold-mode droop considerations.

A two-pole passive RC lowpass bandlimiting and antialiasing filter is described by equations (6-9) and (6-10) and shown in Figure 6-14. This mechanization, which has a rolloff characteristic of -40 dB/decade from its -3-dB f_c, is more efficient in conserving sample rate than a one-pole design in antialiasing applications. However, the one-pole passive RC lowpass filter response of equation (6-11) and (6-12) has a flatter passband characteristic

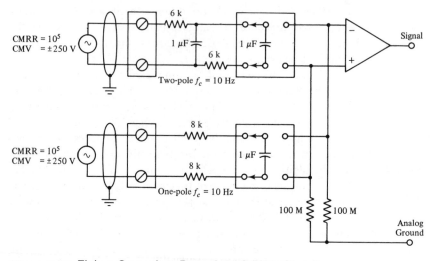

Figure 6-14 Flying-Capacitor Data-Acquisition Circuits

than the two-pole device and therefore exhibits less gain error in the passband. At $0.1 f_c$, for example, gain error for the two-pole RC is 4.5%, whereas for the one-pole RC it is only 0.5%. Consequently, the one-pole filter contributes substantially less amplitude error to the data-acquisition-system error budget. Both filters are suitable primarily for dc and sinusoidal signals, and their RC values must be chosen with the necessary component matching to preserve the incircuit CMRR. These filters also provide normal-mode attenuation for any interference or out-of-band signals above their cutoff frequencies.

Two-pole RC lowpass:

$$f_c = \frac{0.06}{RC} \quad \text{Hz} \qquad (6\text{-}9)$$

$$A_{(f)} = \frac{1}{\sqrt{1 + 9\left(\dfrac{f}{f_c}\right)^2 + \left(\dfrac{f}{f_c}\right)^4}} \qquad (6\text{-}10)$$

One-pole RC lowpass:

$$f_c = \frac{0.159}{RC} \quad \text{Hz} \qquad (6\text{-}11)$$

$$A_{(f)} = \frac{1}{\sqrt{1 + \left(\dfrac{f}{f_c}\right)^2}} \qquad (6\text{-}12)$$

An austere data-acquisition system suitable for low data rates is designed around a voltage-to-frequency converter, as shown in Figure 6-15.

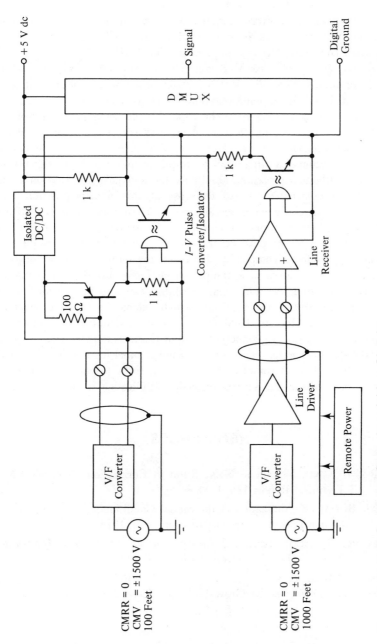

Figure 6-15 V/F Converter Data-Acquisition Circuits

The cost of this system is low because the V/F converter forms the input circuit of a voltage-to-frequency A/D converter. All the basic requirements for a data-acquisition and conversion system are provided, including the capability of locating V/F converters at transducers up to 100 feet from the digitally multiplexed converter. Optical couplers provide CMV isolation of ±1500 V between the analog-encoded V/F converter binary output signals and the multiplexer inputs. Since the V/F converter develops an output that is the average of its input signal, noise rejection also is provided.

Some V/F converters have the capability of being powered from the central location of the multiplexed converter as shown with the current-to-voltage pulse translator circuit. However, V/F converters have no CMRR without the addition of a differential preamplifier, and their input-signal range is unipolar without a preceding absolute-value and polarity-detection circuit. But both these circuits add to the cost of this data-acquisition input arrangement, especially when remote power requirements are considered, thereby diminishing its economic advantage. Consequently, the utility of this method is greatest for unipolar dc transducer signals that do not require this additional input-circuit complexity. However, the location of V/F converters at the transducer source is a compensating factor usually requiring less interference rejection. An additional capability can also be provided by the addition of line drivers and receivers, which extend the signal transmission distance from 100 feet to 1000 feet in noisy environments. The digital multiplexer of Figure 6-15 connects the input channels to gated counter serial-to-parallel A/D converter, remaining on each channel for the duration of the conversion period T. A complete gated-counter, serial-to-parallel V/F−A/D converter suitable for this source-conversion application is shown in Figure 7-23.

REFERENCES

1. R. M. FANO, "Signal to Noise Ratio in Correlation Detectors," MIT Technical Report 186, 1951.
2. M. BUDAI, "Optimization of the Signal Conditioning Channel," Senior Design Project, University of Cincinnati, 1978.
3. H. R. RAEMER, *Statistical Communications Theory and Applications*, Prentice-Hall, Englewood Cliffs, N.J., 1969.
4. M. SCHWARTZ, W. BENNETT, and S. STEIN, *Communications Systems and Techniques*, McGraw-Hill, New York, 1966.
5. P. GARRETT, *Analog Systems for Microprocessors and Minicomputers*, Reston, Reston, Va., 1978.
6. P. GARRETT, "Optimize Transducer/Computer Interfaces," *Electronic Design*, May 24, 1977.

7. M. SCHWARTZ, *Information Transmission Modulation and Noise*, 3rd ed., McGraw-Hill, New York, 1980.

8. R. L. MORRISON, "Getting Transducers to Talk to Digital Computers," *Instruments and Control Systems*, January 1978.

9. J. MILLMAN, *Microelectronics*, McGraw-Hill, New York, 1979.

10. *Linear Databook*, National Semiconductor, Santa Clara, Calif., 1978.

11. *Data Acquisition Products Catalog*, Analog Devices, Norwood, Mass., 1978.

12. E. L. ZUCH, "Principles of Data Acquisition and Conversion," *Digital Design*, May 1979.

PROBLEMS

6-1. A 100-Hz sinusoidal signal in random noise of 6-kHz bandwidth and an SNR of 10 is upgraded by means of linear filtering. Apply the methods of Section 6-2 to determine the input and achievable output amplitude errors for a three-pole Butterworth lowpass filter, observing f_c requirements in accordance with Table 4-9.

6-2. Design a universal signal-conditioning channel consisting of an AD288 isolation instrumentation amplifier with switch-selectable gain values of 1, 10, and 100, where $A_{v_{\text{diff}}} = 1 + 100 \text{ k}/R_G$. This is to be followed by a three-pole unity-gain lowpass active filter with a selectable Butterworth or Bessel characteristic through switchable capacitors. Show the circuit with all components and calculations for gang-switch selectable corner frequencies of 10 Hz, 100 Hz, and 1 kHz, with the lowest-frequency impedance scaled using 100-k resistors.

6-3. Determine the $\text{SNR}_o/\text{SNR}_i$ improvement possible for random interference with the signal-conditioning channel of Problem 6-2 for each gain and filter-cutoff combination possible. Consider the filter efficiency K to be 0.9 and CMRR 10^5. Amplifier f_{hi} versus $A_{v_{\text{diff}}}$ values are: 3.5 kHz—1, 3.3 kHz—10, and 1.5 kHz—100.

6-4. A differential-lag circuit offers useful signal conditioning with a differential input, one-pole RC response, and an operational-amplifier output. Incircuit CMRR is limited principally by the matching of feedback RC products. Determine the signal quality and overall $\epsilon_{\text{channel\%FS}}$ error including appropriate component values for an FET amplifier with 2% component matching. $V_{\text{diff}} = 10 \text{ mV}$ dc, $V_{cm} = 1 \text{ V}$ rms at 60 Hz, $V_o = 1 \text{ V}$, and $f_c = 1 \text{ Hz}$. The amplifier error below consists only of dc uncertainty and does not include its CMRR performance. $\epsilon_{\text{sensor\%FS}} = 0.75\%$, $\epsilon_{\text{ampl\%FS}} = 0.5\%$, and $\epsilon_{\text{filter\%FS}} = 0.25\%$.

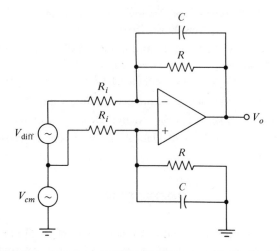

6-5. Compare the amplitude response of one- and two-pole RC lowpass filters using equations (6-10) and (6-12), and determine which is the better choice with regard to passband gain error.

6-6. Compare and diagram eight-bit data-acquisition systems for dc-type signals using the different remote source-conversion approaches of V/F and A/D conversion. Show the required components for each, and choose the method that offers the largest performance/cost ratio.

devices for
data conversion

7-0 INTRODUCTION

Data-conversion systems provide the interface between the continuous-time signals representing measurements of physical parameters and their discrete-time digital equivalent. Today's emphasis on digital systems for industrial control and scientific applications has made analog input/output an important function. This function is performed by analog-to-digital (A/D) and digital-to-analog (D/A) converters and supporting analog multiplexers (AMUX) and sample-hold (S/H) devices. Before we explore the theory and considerations involved in the design of data-conversion systems, it is advantageous to understand the characteristics and various mechanizations of these data-conversion devices. This chapter explains them in detail, including their individual error budgets, and also examines special conversion devices less frequently used.

7-1 ANALOG MULTIPLEXERS

An analog multiplexer (AMUX) is a device that selects one from among available signals according to a digital code. Its primary application is in the front end of data-conversion systems, as illustrated in the preceding chapter.

However, such devices are also used as demultiplexers in data-distribution systems and to select the gain-determining resistors in programmable instrumentation amplifiers. An AMUX consists of several switches connected in parallel at its output controlled by a logic decoder. These bilateral devices enable inputs and outputs to be reversed, thereby permitting signal multiplexing and demultiplexing. The logic is normally designed to open the switches faster than it closes them (break-before-make) to avoid shorting channels together. It is also usual to provide a disable function that turns off all switches to facilitate channel expansion. Figure 7-1 presents the possible interconnection configurations for a multiplexer.

Historically, analog signals have been switched by mechanical devices such as reed relays. These continue to offer advantages in low-data-rate applications requiring high isolation, such as provided by the flying-capacitor multiplexer. Below about 10-mV signal levels mechanical switching is still recommended because of offset-voltage problems. Owing to the speed and

Figure 7-1 Multiplexer Input Connections

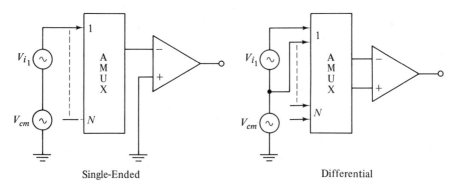

Single-Ended

Differential

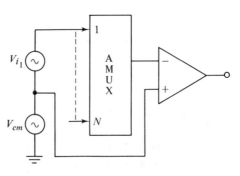

Pseudo-Differential

TABLE 7-1
MULTIPLEXER-SWITCH CHARACTERISTICS

TYPE	ON RESISTANCE	OFF RESISTANCE	SPEED
Reed	0.1 Ω	$10^{14}\,\Omega$	1 ms
CMOS	1.5 kΩ	$10^8\,\Omega$	500 ns
JFET	100 Ω	$10^9\,\Omega$	200 ns

complexity of modern data-acquisition requirements, however, relays have largely been superseded by solid-state switches. Field-effect transistors, both JFET and CMOS, are universally used for electronic multiplexer switches and have replaced bipolar transistors in this application because of the voltage-offset problems of the latter. Junction FET switches have greater electrical device ruggedness and are capable of about the same switching speeds as CMOS devices, but the constant low ON resistance of JFET devices usually eliminates the requirement for an output buffer amplifier. Nevertheless, CMOS switches have become dominant in multiplexer applications primarily because of their unfailing turnoff and their ability to accommodate signal levels up to the supply voltage. A stable ON resistance is achieved with the symmetry provided by paralleled p- and n-channel devices even with varying signal amplitudes. The characteristics of these switch types are summarized by Table 7-1.

Errors associated with analog multiplexers include the static transfer error defined by equation (7-1). This is largely determined by the AMUX contribution to an input voltage-divider effect and is typically 0.01% with an output buffer amplifier which is frequently provided by the sample-hold. Other errors that can be significant are OFF-channel leakage current and the settling-time constant formed by the total AMUX output capacitance with the resistance of the ON channel. Crosstalk between channels may also be of interest. This is typically of the order of -80 dB, or 0.001% error, with presently available devices. CMOS switches exhibit a nominal 1-nA leakage current and 25-pF output capacitance, as described by the CMOS switch equivalent of Figure 7-2. The total leakage current of all OFF channels flows through the source and multiplexer resistances, causing a voltage offset error. With a source resistance of 500 ohms and 15 CMOS OFF channels, this offset totals 30 μV, which is less than 0.001% error with premultiplexer signal conditioning having 10-V_{FS} output scaling.

The switching-time requirement includes that of the actual switch transition plus the settling time to the number of multiplexer output RC time constants of interest. A 2-k total source and multiplexer resistance with 16 channels of 25 pF each provide a 0.8-μs time constant, and 7.2-μs (9RC) are

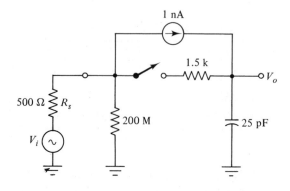

Figure 7-2 CMOS Switch Equivalent Circuit

required for guaranteed settling to 0.012% following a channel transistion involving signal rollover from full-scale positive to full-scale negative values. However, three or four time constants are sufficient for settling within the 1-LSB resolution requirements for eight-bit A/D conversion with 1-V signal changes. CMOS multiplexer transition times add approximately 0.5 μs to this time requirement. Although these parameters may be acceptable for multiplexing 16 channels, expansion to 64 or more would likely impose an unacceptable settling time, necessitating a tiered multiplexer array as suggested by Figure 7-3. Tiered multiplexing produces a reduction in both offset voltage and settling time proportional to the division of channels. Representative analog multiplexer error budgets are provided by Table 7-2, where internal BeO junctions maintain the thermocouple errors of reed switches to the order of 0.001%.

Figure 7-3 Tiered Multiplexer Array

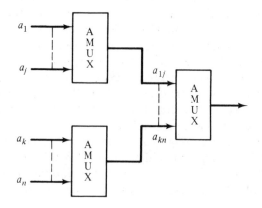

TABLE 7-2
REPRESENTATIVE MULTIPLEXER ERROR BUDGETS

	REED	*CMOS*
Transfer error	0.01%	0.01%
Crosstalk error	0.001	0.001
Leakage error	—	0.001
Thermal offset	0.001	—
RMS total$_{\%FS}$	0.01%	0.01%

$$\text{transfer error} = \frac{V_i - V_o}{V_i} \times 100\% \qquad (7\text{-}1)$$

Figure 7-4 shows a constant-impedance, four-channel JFET multiplexer combined with an operational amplifier to form a programmable gain amplifier (PGA). The p-channel JFET devices may be directly driven with TTL logic levels, with a "1" opening the switches, which simultaneously

Figure 7-4 Programmable-Gain Amplifier

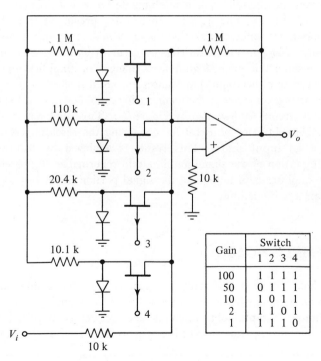

Gain	Switch			
	1	2	3	4
100	1	1	1	1
50	0	1	1	1
10	1	0	1	1
2	1	1	0	1
1	1	1	1	0

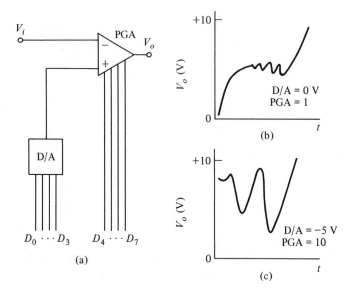

Figure 7-5 Programmable Subtractor Circuit

forward-biases the shunting diodes to eliminate noise and maintain constant-impedance conditions. The JFET switches are prevented from becoming forward-biased to input signals by the ±50-mV clamping action of the amplifier summing junction. Gain changes occur by shunting combinations of feedback resistors; this method minimizes changes in amplifier input uncertainty, because amplifier input impedances are not disturbed. The addition of a digital-to-analog converter to the noninverting amplifier terminal extends the utility of this circuit, as described by Figure 7-5. This combination PGA and programmable subtractor is useful for enhancing the resolution of selectable portions of an input signal. This result is achieved by subtracting an appropriate portion of the signal amplitude to renormalize the baseline, then increasing amplifier gain to expand the signal portion of interest as much as the full-scale amplitude limit.

7-2 SAMPLE-HOLDS

A sample-hold circuit is essentially a signal memory device that can store a voltage on an amplifier-buffered capacitor for periods ranging from microseconds to hours. Sample-hold functions are frequently required in data-conversion and analog signal-processing systems to freeze fast-changing signals during conversion, and for temporary storage in data-distribution and computation circuits. A sample-hold may be used to take a rapid signal sample with the majority of the sample-hold cycle spent in the hold mode, or it can track a signal over most of its cycle and hold it for a short duration. An application for the former would be output-signal distribution, and for the latter, data acquisition where the longer sample time allows virtually complete signal settling.

Sample-holds are available in several circuit variations, each suited to specific speed or accuracy requirements. Figure 7-6 shows a popular circuit that is both fast and accurate. The high-impedance input amplifier is especially useful for multiplexer interfacing, and it also provides an error-correction function during sampling by comparing the output and input voltages and charging the capacitor to reduce this difference to zero. The clamping diodes insure that the input amplifier remains stable by providing a feedback path in the hold mode when the JFET switch is open.

The performance of a sample hold is largely determined by the bandwidth and current limitations of the input amplifier, which reflects its ability to drive the hold capacitor. Amplifier slew rate and settling characteristics for a capacitive load are developed in detail in Section 3-1. In the sample mode, the charge on the capacitor is initially changed at the slew-limited output-current capability I_o of the input amplifier. This is illustrated in Figure 7-7. As the capacitor voltage enters the final settling band coincident with the linear region of amplifier operation, slight ringing may be in evidence, depending upon amplifier stability. The acquisition time determines the maximum sample rate of a sample hold; it is due to both amplifier slew rate (I_o/C) and the resistance of the hold-capacitor charging circuit. The latter consists of the amplifier output resistance R_o and JFET resistance R_{ON}. Equation (7-2) approximates the acquisition time for 0.01% acquisition error, where $|V_o - V_i|$ represents the difference between the sample-hold output and

Figure 7-6 Closed-Loop Sample Hold

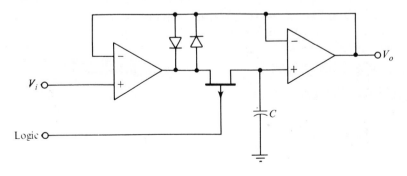

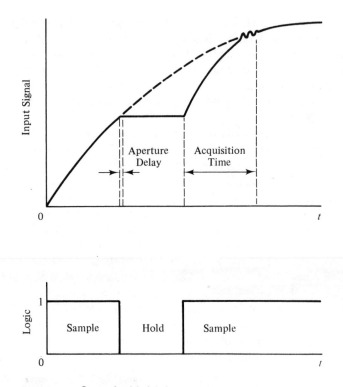

Figure 7-7 Sample-Hold Aperture and Acquisition

the signal to be acquired. Acquisition times range from about 0.1 μs to 10 μs with commercially available sample-hold devices.

$$\text{acquisition time} = \frac{|V_o - V_i|C}{I_o} + 9(R_o + R_{ON})C \quad \text{seconds} \qquad (7\text{-}2)$$

A crucial part of the sample-hold operation, and one that may lead to confusion, is involved with the sample-to-hold transition. At least three time definitions are associated with this interval: (1) aperture delay time, (2) aperture uncertainty time, and (3) aperture time. *Aperture delay time* is simply the elapsed time between issuance of the hold command and the achievement of a high-impedance switch condition. For presently available sample-holds this time ranges from about 1 μs to 1 ns. Closely associated with aperture delay is *aperture uncertainty*, which is a time jitter typically 1% to 10% of the aperture delay. Aperture uncertainty is attributable to the logic circuit and analog switch and can be related to the permissible input signal rate-of-change for signal acquisition to the percent of full-scale of interest. The effect of

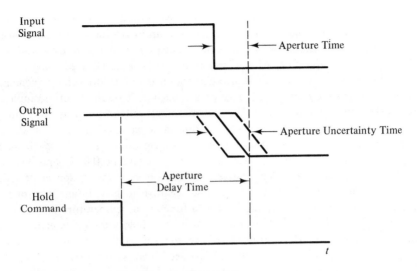

Figure 7-8 Aperture-Time Relationships

aperture uncertainty upon attainable accuracy in data-conversion systems is developed in Section 8-2, where aperture-uncertainty error curves also are presented. Ideally, a sample-hold is capable of taking a point sample of the input signal in an infinitesimal time of switch operation. In reality, owing to both the finite input amplifier bandwidth and switch turnoff time, the input signal is actually averaged to the final acquired value, rather than making an instantaneous transition. This brief averaging interval, or *aperture time*, occurs within the aperture delay time as illustrated in Figure 7-8.

The derivation of a sample-hold error budget consists of essential contributions due to the acquisition time, capacitor droop, dielectric absorption, offset voltage, and hold-mode feedthrough. This error budget is summarized in Table 7-3. Design for 0.01% acquisition error is standard, and the acquisition time necessary to achieve this was presented by equation (7-2). Hold-capacitor voltage droop is determined primarily by the output amplifier bias-current

TABLE 7-3
REPRESENTATIVE SAMPLE-HOLD ERROR BUDGET

Acquisition error	0.01%
Droop $(25\ \mu V/\mu s)(6\ \mu s\ hold)$ in $10\ V_{FS}$	0.0015
Dielectric absorption	0.01
Offset $(50\ \mu V/°C)(20°C)$ in $10\ V_{FS}$	0.01
Hold-jump error	0.001
Feedthrough	0.005
RMS total$_{\%FS}$	0.02%

requirements and by insulation current leakage associated with the capacitor circuit. Hold-mode droop dV/dt decreases with an increasing hold capacitance. However, a larger capacitor increases the acquisition time, providing a tradeoff that results in a typical capacitor choice in the 0.01-μF to 0.001-μF range.

Error from capacitor dielectric absorption is due to the voltage memory creep following the change in charge on a capacitor; it results from incomplete repolarization of the dielectric. We can minimize voltage creep by sustaining the sample mode as long as possible during a sample-hold cycle and by choosing capacitor types with very low dielectric absorption. Polycarbonate capacitors exhibit 0.05% dielectric absorption, polystyrene 0.02%, and Teflon 0.01%. These values are conservative, and the dielectric absorption error will be somewhat lower for long sample times relative to hold times. The offset voltage performance of a sample-hold as a function of temperature is typically in the range of 50 μV/$^{\circ}$C, plus a nominal 0.001% hold-jump offset error from the logic signal, which is transferred by the capacitance of the switch at turnoff. Finally, hold-mode feedthrough is specified as the percentage of an input sinusoidal signal at the output; in well-designed sample-holds it is 0.005%.

Sample-hold devices can benefit even ultrafast video A/D converters by minimizing their signal-acquisition and settling-time requirements. Simultaneous data acquisition utilizing multiple sample-holds is required in some applications, such as seismographic measurements, where all analog signals must be acquired precisely at the same time, as shown in Figure 7-9. For this application data skew will be minimized if the S/H devices are matched in their bandwidths and aperture delay times and are chosen for very small aperture uncertainty times. Ultrafast S/H devices are also utilized to deglitch

Figure 7-9 Simultaneous Data Acquisition

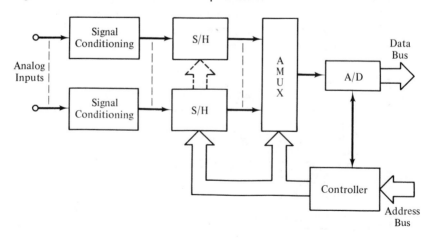

circuits that have expected transients by sequencing the sample-hold to be in the hold mode during the transient interval.

7-3 DIGITAL-TO-ANALOG CONVERTERS

The basic structure of all conventional D/A converters (DAC's) involves a network of precision resistors, a set of switches, some form of voltage scaling to adapt the switch drives to the specified logic levels, and a voltage reference, which is usually temperature-compensated. Each switch closure adds a binary-weighted current increment to an output summing bus. A current-output DAC, provided by omitting the current-to-voltage-conversion operational amplifier, offers higher speed in the amount of settling time saved by omitting this amplifier. For the weighted-resistor DAC of Figure 7-10, a 12-bit converter requires a range of resistance values of 4096-to-1 with an LSB value of perhaps 41 M. The use of repeated resistance values in the R-$2R$ network DAC of Figure 7-11 has become almost universal because of the efficiency

Figure 7-10 Weighted Resistor D/A Converter

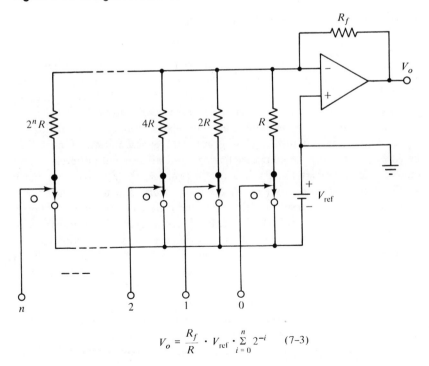

$$V_o = \frac{R_f}{R} \cdot V_{\text{ref}} \cdot \sum_{i=0}^{n} 2^{-i} \quad (7\text{-}3)$$

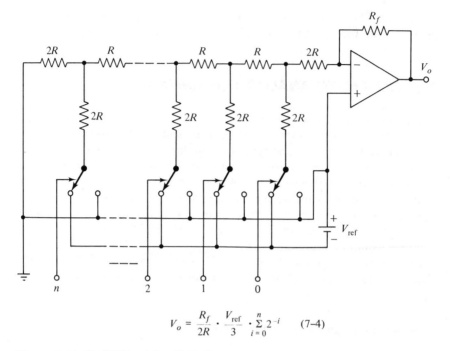

$$V_o = \frac{R_f}{2R} \cdot \frac{V_{ref}}{3} \cdot \sum_{i=0}^{n} 2^{-i} \quad (7\text{-}4)$$

Figure 7-11 $R-2R$ Resistor D/A Converter

associated with the limited number of required resistance values. This mechanization has also permitted the realization of accurate, low-cost monolithic DAC's. Settling times for typical 12-bit D/A converters are typically 0.5 μs for current output and 5 μs for voltage output.

A descriptive way of indicating the relationship between analog and digital conversion quantities is a graphical representation. Figure 7-12 describes a three-bit D/A converter having eight discrete output levels ranging from zero to seven-eighths of full scale. In practice, the zero bar may not be exactly zero because of offset error, the range from zero to 7/8 may not be precisely encoded as a result of gain error, and differences in the heights of the bars may not change uniformly because of nonlinearity. Nonlinearity is the most difficult error to compensate for because it cannot be eliminated by adjustment. Differential nonlinearity and its variation with temperature is critical, however, because it indicates the difference between actual output-voltage changes and the ideal voltage change for 1-LSB code increments. For example, a DAC with a $1\frac{1}{2}$ LSB equivalent output-voltage change for a 1-LSB code change exhibits $\frac{1}{2}$LSB differential nonlinearity. Differential nonlinearity may be twice an LSB code change, but a D/A converter is said to be nonmonotonic if this exceeds 1 LSB, making it no longer single valued.

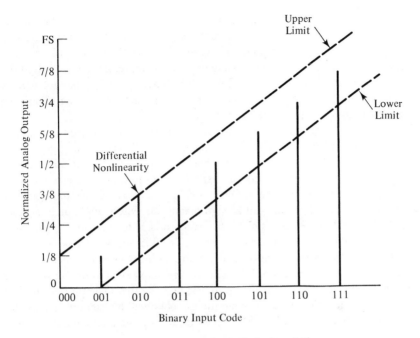

Figure 7-12 Three-Bit D/A Converter Relationships

Offset error is the output voltage of a DAC with a zero input code and is attributable to the output summing amplifier. Offset can be zeroed, but offset drift associated with both the amplifier and reference cannot be eliminated by adjustment. *Gain error* is the departure from the design output voltage for a given input code. This error is usually due to variance in the ladder resistance values, output amplifier, or reference. Table 7-4 summarizes these errors with an assumed 20°C temperature variation for a typical 10-V_{FS} 12-bit D/A converter. D/A converters differ in a number of ways with regard to their input data. First, the coding must be appropriate and its sense understood (positive-true, negative-true). The number of bits wordlength must also be

TABLE 7-4
REPRESENTATIVE 12-BIT DAC ERROR BUDGET

Differential nonlinearity ($\pm\frac{1}{2}$ LSB)	0.012%
Linearity tempco (2 ppm/°C)(20°C)	0.004
Gain tempco (20 ppm/°C)(20°C)	0.040
Offset tempco (5 ppm/°C)(20°C)	0.010
RMS total$_{\%FS}$	0.04%

sufficient for the output actuation task. For example, eight bits may be entirely adequate for an analog controller remote-setpoint input driven from the I/O system of a supervisory computer. However, the wordlength of a 16-bit DAC could be important to provide smooth output waveforms for a digitally synthesized function generator. The D/A converter error budget can be minimized in all applications by the choice of a converter having about two bits more than the required binary resolution.

Frequently used input codes for D/A converters are the sign-magnitude, offset-binary, and two's-complement bipolar codes illustrated in Figure 7-13. Sign magnitude, a straightforward method for expressing digital quantities in terms of an analog equivalent, is especially useful for outputs that are frequently in the vicinity of zero. Offset binary is the easiest code to implement with converter circuitry and, unlike sign magnitude, has a single unambiguous code for zero. However, a major bit transition occurs for the zero output code, making it somewhat susceptible to errors in this region. Offset binary is readily converted to the more computer-compatible two's-complement code by simply complementing the MSB. The two's-complement code consists of a binary number for the positive magnitudes, and the two's complement of each positive code for the negative magnitudes, obtained by complementing the binary number and adding 1 LSB. Note that the sign-magnitude code exhibits

Figure 7-13 Three-Bit DAC Input Codes

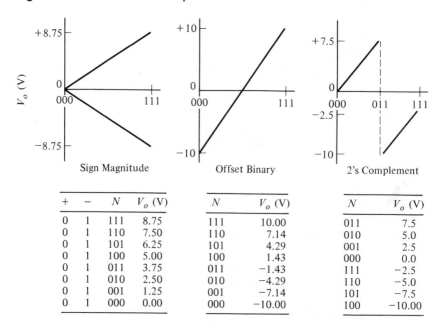

+	−	N	V_o (V)
0	1	111	8.75
0	1	110	7.50
0	1	101	6.25
0	1	100	5.00
0	1	011	3.75
0	1	010	2.50
0	1	001	1.25
0	1	000	0.00

N	V_o (V)
111	10.00
110	7.14
101	4.29
100	1.43
011	−1.43
010	−4.29
001	−7.14
000	−10.00

N	V_o (V)
011	7.5
010	5.0
001	2.5
000	0.0
111	−2.5
110	−5.0
101	−7.5
100	−10.00

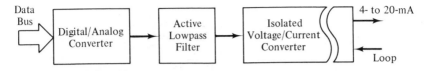

Figure 7-14 Analog Output Channel

twice the resolution of the other two codes but requires a fourth sign bit in its implementation. By omitting this sign bit we obtain a unipolar binary code.

A D/A converter can be considered a digitally controlled potentiometer that provides an output current or voltage normalized to its full-scale value. Multiplying D/A converters differ from conventional DAC's in that an external reference voltage is utilized as the second variable. If the reference voltage can assume bipolar values and the digital input can also represent bipolar polarities, then four-quadrant multiplication is possible. If either is restricted to unipolar operation, then only two-quadrant multiplication is realizable. Multiplying DAC's are useful for implementing digitally controlled attenuators and programmable-gain amplifiers, the latter by including the D/A converter in the feedback loop of an operational amplifier. Conventional DAC's are utilized in the mechanization of successive approximation analog-to-digital converters to generate an input comparator analog reference signal from the converter binary output as described in the following section. Figure 7-14 describes a frequent application for D/A converters in the data-distribution channel of a process automation computer interface. This implementation is developed in detail in Chapter 9.

7-4   ANALOG-TO-DIGITAL CONVERTERS

A/D converters are the basic interfaces between the world of analog parameters and digital signals. Techniques that are commonly applied and commercially available are of the six types shown in Table 7-5, classified as either integrating or voltage-comparison methods. Other implementations are possible and have been utilized in the past, such as single-slope and pulse-width converters, but these have been superseded by the presently available devices. All A/D converters implement the operations of sampling, quantizing, and encoding illustrated by Figure 7-15 in order to relate analog and digital values. The detailed considerations associated with these functions required in the design of data-conversion systems are developed in Chapter 8. One conversion is performed each period T, whereby a numerical value derived from the quantizing levels is converted to the appropriate output code.

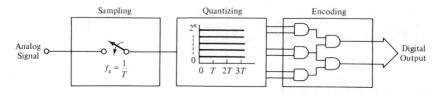

Figure 7-15 A/D Conversion Operations

Quantization of a sampled analog waveform involves the assignment of a finite number of amplitude levels increasing to some full-scale (FS) value such as 10 V. The quantizing interval q shown in Figure 7-16(a) represents the converter least significant bit (LSB) resolution limit; its value is determined

Figure 7-16 Three-Bit A/D Converter Relationships: (a) Quantizing Interval, (b) Quantization Error

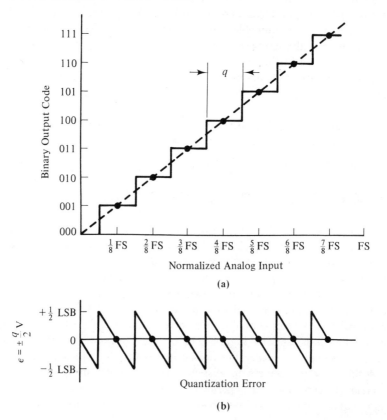

TABLE 7-5
A/D CONVERTER METHODS

METHOD AND SPEED T	SIGNALS	TECHNIQUE	12-BIT RATE $\left(\frac{1}{T}\right)$	8-BIT RATE $\left(\frac{1}{T}\right)$
Integrating, 1 s to 1 ms	DC and sinusoidal	Dual slope	100 Hz	1 kHz
		Charge balancing	50 Hz	500 Hz
		Voltage to frequency	2.5 Hz	40 Hz
Voltage comparison 1 ms to 10 ns	DC, sinusoidal, and complex harmonic	Successive approximation	100 kHz	1 MHz
		Tracking	250 Hz	4 kHz
		Simultaneous	—	35 MHz

TABLE 7-6
12-BIT A/D CONVERTER OUTPUT CODES

UNIPOLAR CODING			BIPOLAR CODING		
FS Fraction	FS (10 V)	Straight Binary	FS Fraction	FS (±5 V)	Two's Complement
FS − 1 LSB	9.99 V	1111 1111 1111	FS − 1 LSB	+4.99 V	0111 1111 1111
$\frac{3}{4}$ FS	7.50 V	1100 0000 0000	$\frac{1}{2}$ FS	+2.50 V	0100 0000 0000
$\frac{1}{2}$ FS	5.00 V	1000 0000 0000	$\frac{1}{4}$ FS	+1.25 V	0010 0000 0000
$\frac{1}{4}$ FS	2.50 V	0100 0000 0000	0	0.00 V	0000 0000 0000
$\frac{1}{8}$ FS	1.25 V	0010 0000 0000	$-\frac{1}{4}$ FS	−1.25 V	1110 0000 0000
1 LSB	0.0024 V	0000 0000 0001	$-\frac{1}{2}$ FS	−2.50 V	1100 0000 0000
0	0.00 V	0000 0000 0000	−FS + 1 LSB	−4.99 V	1000 0000 0001

both by the full-scale amplitude value and by the total number of levels 2^n provided by an n-bit converter. With the dominant uniform quantizing algorithm, the quantization error ϵ is a sawtooth function ranging in value up to $\pm q/2$ volts ($\pm \frac{1}{2}$ LSB) as shown in Figure 7-16(b) regardless of the A/D converter method or implementation. The straight binary and two's-complement codes most frequently used with A/D converters are presented in Table 7-6. The MSB has a weight of $\frac{1}{2}$, and so on to the LSB, which has a weight of $2^{-n} \cdot$ FS. Note that the maximum output code does not correspond to the full-scale value, but to $(1 - 2^{-n}) \cdot$ FS.

Integrating converters provide noise rejection for the input signal at an attenuation rate of -20 dB/decade of frequency, with practically achievable nulls to -60 dB at multiples of the integration period. The ability of an integrator to provide this response can be determined by examination of its impulse response $h(t)$ and frequency-domain response $H(\omega)$, provided by the integral and Fourier expressions of equations (7-5) and (7-6). These are illustrated in Figure 7-17. Note that this noise rejection is not provided when a sample-hold precedes the A/D converter. The more commonly applied voltage-comparison A/D method generally provides increased speed over the integrating method but does not offer a noise-rejection capability. Special conversion techniques are less frequently used because of their specific features. For example, the ultrahigh-speed simultaneous converter is for video and radar data, and the slower voltage-to-frequency A/D converter is especially useful for inexpensive, remote-source-conversion systems as illustrated in Figure 6-15. For most converters the conversion rate is decreased by an order of magnitude between eight and 12 bits, as described by Table 7-5, where these rates correspond to conversion of full-scale input signals.

The successive-approximation technique is popular because it is capable of high resolution, is fast, and has a fixed conversion time independent of the input-signal amplitude. The latter is advantageous for application in multiplexed data-conversion systems for computer input. This converter operates by comparing the output of an internal D/A converter with the input signal, one bit at a time, as illustrated in Figure 7-18. Therefore, n, fixed time periods are needed to deliver an n-bit output providing a constant conversion time. Conversion accuracy depends upon the stability of the reference voltage and analog gain and offset errors as with the D/A converter of the previous section. Its error budget also includes a term representing the $\pm \frac{1}{2}$-LSB quantization error as tabulated in Table 7-7. Successive-approximation converters are available in monolithic, hybrid, and discrete-circuit modular implementations with attendant cost-performance tradeoffs.

Dual-slope integrating converters operate by the indirect method of converting a voltage to a time period that is then totaled by a counter. Besides the input-averaging capability possessed by integrating converters, dual-slope conversion has other advantages. For one, it is self-calibrating in operation,

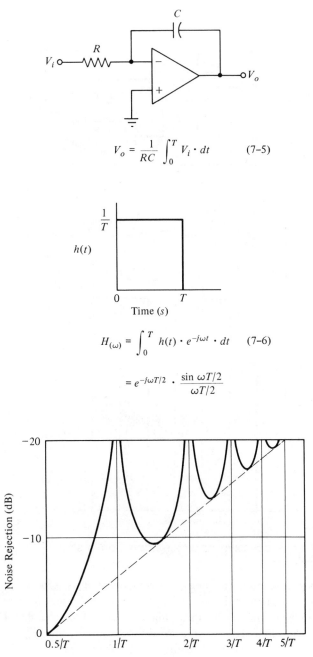

$$V_o = \frac{1}{RC} \int_0^T V_i \cdot dt \qquad (7\text{-}5)$$

$$H_{(\omega)} = \int_0^T h(t) \cdot e^{-j\omega t} \cdot dt \qquad (7\text{-}6)$$

$$= e^{-j\omega T/2} \cdot \frac{\sin \omega T/2}{\omega T/2}$$

Figure 7-17 Integrating-Converter Noise Rejection

179

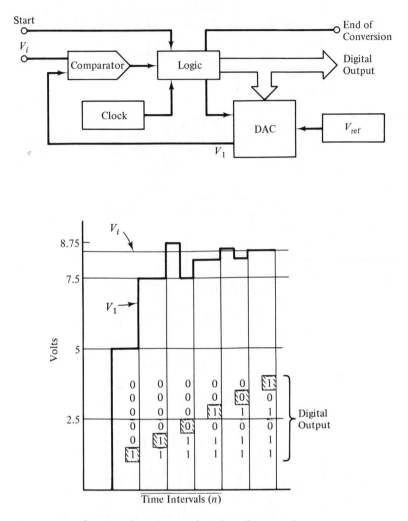

Figure 7-18 Successive-Approximation Conversion

which makes it immune to long-term component drift. Operation occurs in three steps, as shown in Figure 7-19. The first is the auto-zero step, which stores the analog errors on the integrator with the input grounded. During the second step the input signal is integrated for a constant time T_1, providing an integrator output proportional to the input voltage. These relationships are described by equations (7-7) and (7-8). In the final step the input is connected to a reference voltage of opposite polarity, and integration proceeds to zero. The clock pulses counted during this variable time T_2 provide a digital measure of the input signal, and the total conversion time $(T_1 + T_2)$ is

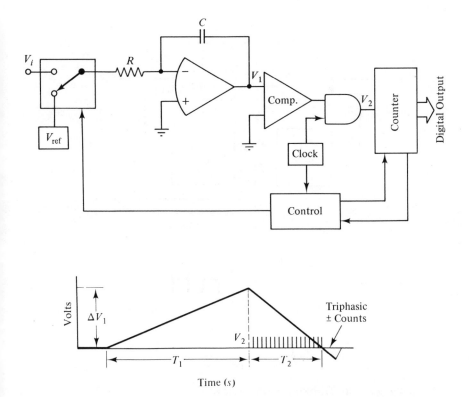

Figure 7-19 Dual-Slope Conversion

$$\Delta V_1 = \frac{1}{RC} \cdot V_i \cdot T_{1_{\text{constant}}} \qquad (7\text{-}7)$$

$$= \frac{1}{RC} \cdot V_{\text{ref}} \cdot T_{2_{\text{variable}}}$$

$$T_2 = \frac{V_i \cdot T_1}{V_{\text{ref}}} \qquad (7\text{-}8)$$

proportional to V_i. Differential linearity is excellent because the output code is generated by a clock and counter. The error budgets developed in Table 7-7 for successive-approximation and dual-slope converters, from a compilation of manufacturers' data, indicate comparable performance over a 20°C temperature variation. Error budgets for other converter types can readily be developed from manufacturers' literature. Commercial dual-slope converters have additional features, including provision for handling positive and negative input signals, and frequently a triphasic third integration for improved accuracy about the zero endpoint.

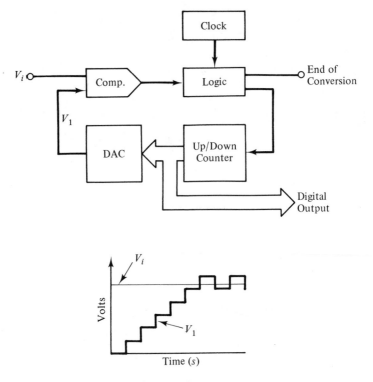

Figure 7-20 Tracking Conversion

The voltage-comparison tracking converter is especially useful for direct-conversion parallel systems that employ one A/D converter per channel. The free-running mechanization described by Figure 7-20 has simple control requirements, and performance is not improved with an S/H device. In operation its conversion time is variable, and it tracks the input signal at a rate of 2^{-n} per count of the n-bit counter. For example, an eight-bit converter with an integral 1-MHz clock will track an input signal up to a $2^{-8} \cdot 1$ MHz, or 4-kHz conversion rate, with 256 counts per conversion. The output state of the comparator determines the direction of the count (up or down). Starting from zero, the converter counts clock pulses that are converted to the analog voltage V_1, which reverses direction when it exceeds V_i. This technique is slower than successive-approximation conversion, because the tracking converter must go through each quantizing level to determine when V_1 equals V_i, whereas the former only goes through the number of bits of the converter. The decision time required to evaluate 1 LSB determines the aperture of this converter; in this example it is approximately equal to one clock cycle or 1 μs.

TABLE 7-7
REPRESENTATIVE A/D-CONVERTER ERROR BUDGETS

12-BIT DUAL SLOPE	
Quantization error $\left(\pm\dfrac{1}{2} \text{ LSB}\right)$	0.012%
Differential nonlinearity $\left(\pm\dfrac{1}{2} \text{ LSB}\right)$	0.012
Gain tempco (25 ppm/°C) (20°C)	0.050
Offset tempco (2 ppm/°C) (20°C)	0.004
Long-term change	0.025
RMS total$_{\%FS}$	0.06%

12-BIT SUCCESSIVE APPROXIMATION	
Quantization error $\left(\pm\dfrac{1}{2} \text{ LSB}\right)$	0.012%
Differential nonlinearity $\left(\pm\dfrac{1}{2} \text{ LSB}\right)$	0.012
Linearity tempco (2 ppm/°C) (20°C)	0.004
Gain tempco (20 ppm/°C) (20°C)	0.040
Offset tempco (5 ppm/°C) (20°C)	0.010
Long term change	0.050
RMS total$_{\%FS}$	0.07%

The integrating charge-balancing A/D converter utilizes a special voltage-to-frequency converter, which converts the input signal to a current I_i and has subtracted from it a reference current I_{ref} greater than the input current for allowable input signals. This difference current is integrated for successive alternations, reversal being determined in one direction by a threshold crossing detector and in the other by one clock count. The conversion period is a constant for this converter, but the number of counts received by the output counter varies. From Figure 7-21 we see that the rate of accumulation of these pulses is directly proportional to the input-signal amplitude. The charge-balancing converter is similar in speed and complexity to the dual-slope converter and is frequently confused with the voltage-to-frequency A/D conversion technique, which is substantially slower. These converters also are available in monolithic form, which makes them economically attractive. Integrating A/D converters generally are not compatible for use with complex harmonic signals, however, because of their conversion speed limitations and signal-averaging characteristic.

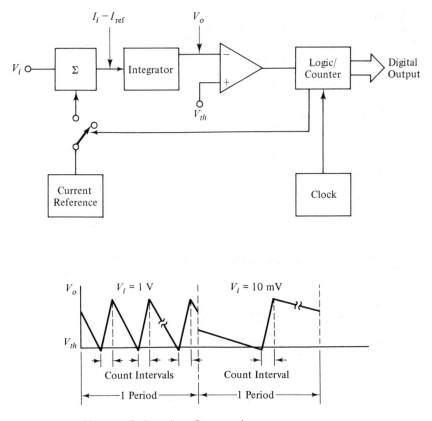

Figure 7-21 Charge-Balancing Conversion

7-5 SPECIAL CONVERTER MECHANIZATIONS

Special converter mechanizations accommodate the requirements of specific but less frequently encountered applications, such as wide signal bandwidths and high-resolution conversion. The representative V/F and F/V converter circuits shown in Figure 7-22 respectively produce and accept constant-width pulses at a rate corresponding to the analog-signal amplitude. These devices may be used together to provide a binary-encoded analog-signal transmission channel offering substantial improvement in signal quality over the transmission of an analog signal through a high-interference environment. They are unipolar devices with linearity to 0.01% and corresponding frequency ranges available to 1 MHz for a six-decade current range (1 nA to 1 mA) and four-decade

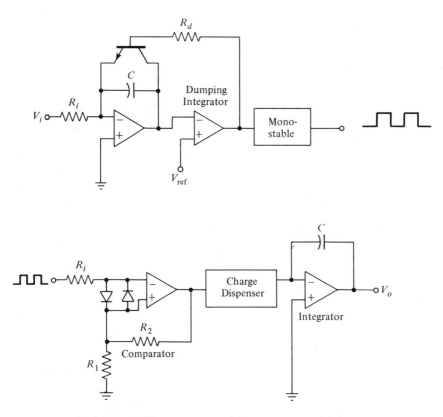

Figure 7-22 Voltage-to-Frequency and Frequency-to-Voltage Converters

voltage range (1 mV to 10 V) of signal amplitude representation. V/F converter bandwidth is variable and proportional to the input-signal amplitude, as related to its output-frequency equivalent. F/V converter bandwidth is primarily determined by its output integrator capacitor.

Voltage-to-frequency A/D conversion has utility principally as an economical, low-data-rate, direct-source-conversion system such as illustrated in Figure 6-15. Figure 7-23 shows V/F-A/D conversion that is a feedforward conversion of input-signal amplitude to a proportional output-pulse repetition rate. These pulses are then counted for a fixed gating time, dependent upon the V/F full-scale equivalent output frequency, in a binary counter. This also results in an averaging of the input signal over this gating-time conversion period for noise rejection equivalent to that of an integrating converter. A conversion period of 25.6 ms is required for eight-bit conversion with a 10-kHz V/F converter, and 0.41 s for 12-bit conversion.

Simultaneous A/D conversion employs an input quantizer composed of

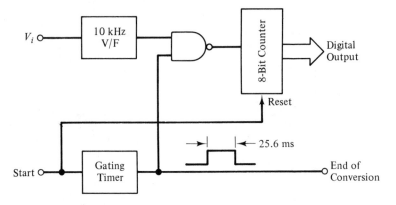

Figure 7-23 Voltage-to-Frequency A/D Conversion

$2^n - 1$ comparators biased 1 LSB apart by a voltage reference. The quantization process is accomplished in the switching time of the comparators; however, an output encoder is required to realize the binary output code of interest, as illustrated in Figure 7-24. Commercial simultaneous converters are available with a 35-MHz conversion rate to eight bits, such as provided by the monolithic TRW 1007 device. A variation of this high-speed A/D conversion technique is the combination of two or more staged four-bit simultaneous converters, each successively converting the difference between the input signal and the analog equivalent of the preceding converter output. This arrangement trades speed for additional resolution, overcoming the simultaneous converter

Figure 7-24 Simultaneous Conversion

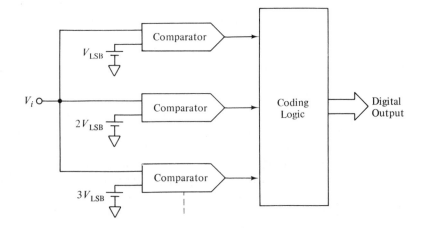

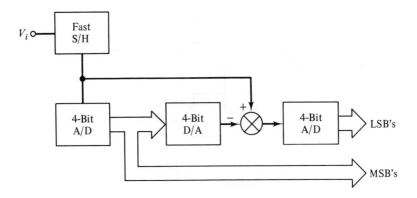

Figure 7-25 Two-Stage Simultaneous Conversion

geometrical increase in the number of comparators required with increasing output binary wordlength. An example of this mechanization, shown in Figure 7-25, is commercially available as the 20-MHz eight-bit Datel-Intersil TV8B converter.

Some applications require digitizing analog signals over a very wide dynamic range. Consider a 10-V to 1-mV data-conversion input-amplitude variation with a full-scale 13-bit resolution requirement, which corresponds to a 1.2-mV LSB. This 8,192-to-1 (2^{13}) dynamic range is not easily realized with conventional uniform quantizing A/D converters because of the associated errors and temperature coefficients. However, a solution is to employ logarithmic compression to effectively increase the gain at minimum input-signal levels prior to application to a conventional A/D converter. This technique is developed in detail in Chapter 2 where it is shown that a four-decade compressor can provide 13-bit resolution represented by eight log bits, as illustrated in Figure 7-26. A commercial embodiment of this technique is available with the Analogic AN8020L logarithmic A/D converter, which performs the log transform digitally and is capable of a 1,000,000-to-1, 20-bit dynamic range. Note that logarithmic compression maintains a constant signal resolution throughout the dynamic range at the expense of high resolution at any point within this range.

Figure 7-26 Thirteen-Bit Resolution Logarithmic Conversion

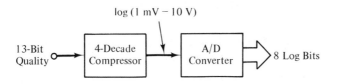

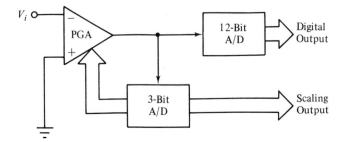

Figure 7-27 Autoranging A/D Conversion

An alternate technique for accommodating wide-dynamic-range signals, which are common in multiplexed low-level transducer systems, is to utilize an autoranging A/D converter whose mechanization is described by Figure 7-27. A fast-settling programmable-gain amplifier (PGA) whose gain is selectable in eight binary multiples between 1 and 128 can be represented by three binary bits. Consequently, a very fast three-bit simultaneous A/D converter can automatically adjust the gain of the PGA so that the input signal is digitized within the practical resolution range of a 12-bit converter while accommodating the dynamic range of interest. Considering a 10-V to 10-mV input signal range and a 16-bit resolution requirement, for example, a 10-mV signal results in the maximum gain of 128, providing the lower six bits of resolution with negligible uncertainty. This appears between output bits 4 (625 mV) and 9 (19.6 mV), which is referenced by the gain scaling output. Unity-gain scaling is provided for input signals between 10 V and 625 mV. A commercial version of this technique is available from Micro Networks Corporation as their MN5410 converter.

The interfacing of electromechanical hardware to digital processors is a frequent requirement in industrial and military systems that require the measurement of angular position. Natural binary is the most commonly used method of representing angles in digital form, where the MSB represents $180°$, the next bit $90°$, and so on as shown in Table 7-8. Digital conversion of a synchro input plus reference signal begins with the Scott-T transformer output signals V_A and V_B of equations (7-9) and (7-10), which are multiplied by the quadrature sin ϕ and cos ϕ functions. These two signals are then subtracted to develop the signals V_E and V_D of equations (7-11) and (7-12), which are in a closed-loop circuit that nulls sin $(\theta - \phi)$ by means of a phase detector and voltage-controlled oscillator (VCO) to provide a digital output ϕ equalling the synchro angle θ. Analog Devices Incorporated offers a series of these conversion products including digital sine/cosine converters, synchro-to-linear converters, and digital vector converters. Synchro-to-digital conversion is shown by Figure 7-28.

TABLE 7-8
BINARY ANGLE REPRESENTATION

BIT	DEGREES
1	180
2	90
3	45
4	22.5
5	11.25
6	5.625
7	2.812
8	1.406
9	0.703
10	0.351
11	0.176
12	0.088

$$V_A = \sin(377t) \cdot \sin\theta \cdot \cos\phi \qquad (7\text{-}9)$$

$$V_B = \sin(377t) \cdot \cos\theta \cdot \sin\phi \qquad (7\text{-}10)$$

$$V_E = \sin(377t) \cdot \sin(\theta - \phi) \qquad (7\text{-}11)$$

$$V_D = \sin(\theta - \phi) \qquad (7\text{-}12)$$

Figure 7-28 Synchro-to-Digital Conversion

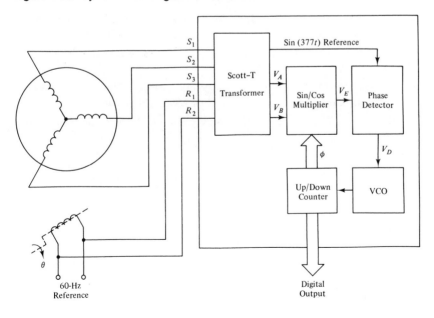

Ratiometric A/D conversion is also occasionally useful in transducer applications that employ an external reference. This mechanization uses a multiplying DAC in the successive-approximation converter of Figure 7-18, with V_{ref} providing both the analog input to this DAC and the transducer excitation, such as required by a bridge circuit. Consequently, the output of the conversion process is always interpreted in terms of the full-scale transducer output regardless of sensor off-ground potentials.

REFERENCES

1. *Analog-Digital Conversion Handbook,* Analog Devices, Norwood, Mass., 1972.

2. E. R. HNATEK, *A User's Handbook of D/A and A/D converters,* John Wiley, New York, 1976.

3. B. M. GORDON, *The Analogic Data-Conversion Systems Digest,* Analogic, Audubon Road, Wakefield, Mass., 1977.

4. E. ZUCH, *Data Acquisition and Conversion Handbook,* Datel-Intersil, 11 Cabot Boulevard, Mansfield, Mass., 1977.

5. D. STANTUCCI, "Data Acquisition Can Falter Unless Components Are Well Understood," *Electronics,* November 27, 1975.

6. B. M. GORDON, "The ABC's of A/D and D/A Converter Specifications," *Electronic Design News,* August 1972.

7. M. LINDHEIMER, "Guidelines for Digital-to-Analog Converter Applications," *Electronic Equipment Engineering,* September 1970.

8. E. ZUCH, "Consider V/F Converters," *Electronic Design,* November 22, 1976.

9. R. ALLEN, "A/D and D/A Converters: Bridging the Analog World to the Computer," *Electronic Design News,* February 5, 1973.

10. E. ZUCH, "Put Video A/D Converters to Work," *Electronic Design,* August 22, 1978.

8

data conversion
systems

8-0   INTRODUCTION

The design of data-conversion systems is largely an analog task. Market pressures increasingly require that such systems provide converted data only to the necessary accuracy at minimum cost. It is essential, therefore, to develop a unified method for the design of these systems in order to accommodate both the component errors and the many interrelationships associated with their assembly into systems. Data conversion is of significant interest for audio and video waveform encoding as well as measurement and control applications.

The first two sections of this chapter explore the theoretical aspects of sampled-data systems including aliasing, quantizing, $\sin x/x$, and aperture errors. The measurement signal type, antialiasing filter parameters, and sampling are coordinated, and the sample-rate requirements for the A/D converted binary resolution of interest are presented. The third section brings together both data acquisition and conversion in the development of detailed error budgets for analog input systems to determine their measurement accuracy. These examples then permit examination of the sensitivities and competing tradeoffs associated with the design of complete systems.

The final section develops the organization of control circuits for both parallel and multiplexed data-conversion systems, including methods for transferring binary data to the digital processor. Conversion-system timing, interfacing options, and control requirements are also explored.

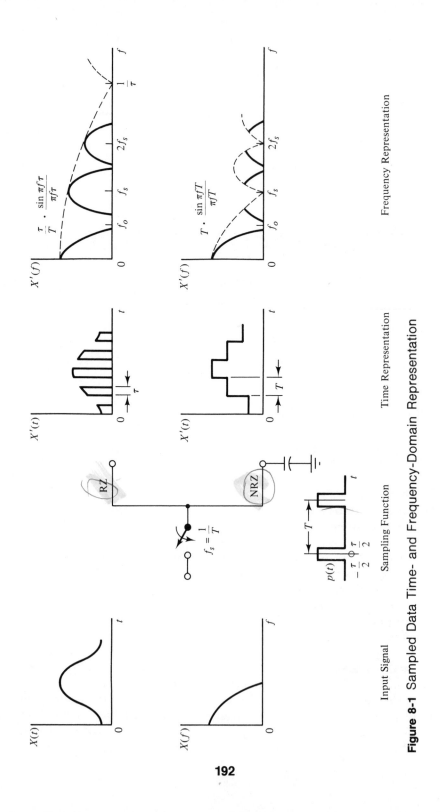

Figure 8-1 Sampled Data Time- and Frequency-Domain Representation

192

8-1 SAMPLE RATE AND ALIASING

The process of analog-to-digital conversion involves the three distinct operations of sampling, quantizing, and encoding. We shall examine these operations in some detail for the insight they provide into data-conversion systems. Sampling can be performed either by a sample-hold or A/D converter that provides nonreturn-to-zero (NRZ) sampling of the input signal, or by a multiplexer switch that imposes return-to-zero (RZ) sampling. Examination of Figure 8-1 illustrates the significant difference between these methods, which are represented in both the time and frequency domains.

A significant consideration imposed upon the sampling operation results from the finite width τ of practical functions. This is shown in Figure 8-1 by the periodic sampling pulses $p(t)$, and it has the important spectral property defined by equation (8-1). Consequently, the product of $p(t)$ of unit amplitude with the input signal $X(t)$ provides the sampled-data signal $X'(t)$, and its spectrum $X'(f)$, determined from the Fourier operations of equations (8-3) and (8-4). It is apparent from this result that the spectrum of a sampled signal consists of its original baseband spectrum $X(f)$ plus an infinite number of images of the original spectrum. These image spectrums are shifted in frequency by amounts equal to the sampling frequency f_s and its harmonics nf_s. The amplitudes of these spectrums are also multiplied by the superimposed $\sin x/x$ functions of equations (8-1) and (8-2), which determine the rate at which these spectrums are attenuated with increasing frequency. The sampled-data bandwidth requirement for RZ sampling is normally considered to be the first null of its $\sin x/x$ function, or $1/\tau$, which includes the majority of the energy of the original continuous signal $X(t)$ in these translated sampled-data spectrums.

Analysis of NRZ sampling can be achieved by examination of the functions shown in Figure 8-2. This frequency-domain representation includes summation of the input signal spectrum $X(f)$ with an inverted $-X(f)$ delayed in time by T, using the pure time-delay operator $e^{-j2\pi fT}$. This summation is then integrated in the frequency domain by the $1/j$ operator resulting in the

Figure 8-2 Equivalent NRZ Frequency Transformation

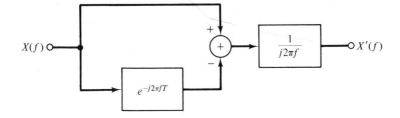

NRZ sin x/x function. Solution of equations (8-2) through (8-4) using this function provides a sampled-data translated spectrum $X'(f)$ with amplitude nulls occurring at f_s and its harmonics. In summary, the sampled-data bandwidth requirement for NRZ sampling is normally considered to be the first null of its sin x/x function, or f_s, and therefore it is more efficient in bandwidth utilization than the $1/\tau$ null with RZ sampling. RZ sampling is used primarily to permit the multiplexing of multiple signals into a single channel, and it has been widely applied in satellite communication systems. NRZ sampling is inherent in the operation of a sample-hold and conventional A/D converter that has an output holding register; for this reason it is universally encountered in computer input systems. Further, the folding frequency for the image spectrums in Figure 8-1 is always located at one-half the sampling rate, or $f_0/2$.

Equation (8-1) provides that the dc component of RZ sampling has an amplitude of τ/T, its average value or sampling duty cycle, which may be scaled as required by the system gain. For the more prevalent NRZ sampling associated with data-conversion devices the dc component of the frequency spectrum is shown in equation (8-2) to be proportional to the sample period T, or $1/f_s$. In practice, however, this gain factor is accommodated in the implementation of data-conversion devices such that A/D-converted V_{FS} input signals can be recovered V_{FS} at a D/A converter output up to the conversion rate at which their specific mechanization permits them to switch and settle. This is effected as a result of the $1/T$ impulse-amplitude response of these networks, which dimensionally normalizes the gain factor to unity at dc.

An important consideration in sampled-data systems is the amplitude attenuation introduced as a result of the convolution of these sin x/x functions with the signal spectrums, illustrated in Figure 8-1, where the amplitude of the composite spectrum is the product of the signal amplitude and the envelope of these functions. The average amplitude error expressed as a percent departure from unity gain is given by equations (8-5) and (8-6). For the more prevalent NRZ sampling, this error is obtained in terms of the required signal BW and chosen sampling rate f_s. It is also useful to note that the specific NRZ sin x/x attenuation at the folding frequency f_0 is always 0.636, or -4 dB.

The data of Table 8-1 are assembled from Tables 1-2, 4-9, and 6-2 to provide the coordination required between the lowpass antialiasing filters and sample rates that reduce signal aliasing to a nominal value. Aliasing is illustrated in the time domain by Figure 8-3, which demonstrates that the unique mapping of a signal to its sampled-data representation does not have a unique reverse mapping. However, an understanding of aliasing can be better accommodated in the frequency domain. This essentially reduces to maintaining an overlap amplitude at the folding frequency f_0 shown in Figure 8-1 of the order of 1% of the passband value. One percent amplitude represents a conservative value because of the normally diminished signal in this frequency region. Aliasing errors in the sampled data are due to downward frequency

TABLE 8-1
SIGNAL, FILTER, AND SAMPLING COORDINATION FOR NOMINAL ALIASING

INPUT SIGNAL		FILTER ORDER (POLES)			SAMPLE RATE	A/D CONVERTER
Type	BW (Hz)	Butterworth ($f_c = 1.6\ BW$)	Besselworth ($f_c = 1.6\ BW$)	Bessel ($f_c = 2.5\ BW$)	f_s	Method
DC and sinusoidal	Fundamental	7			$4f_c$	Integrating or voltage comparison
		6			$4f_c$	
		5			$5f_c$	
		4			$6f_c$	
		3			$8f_c$	
		2			$16f_c$	
		1			$96f_c$	
Complex harmonic	10 Fundamental		4	7	$6f_c$	Voltage comparison
				6	$7f_c$	
				5	$8f_c$	
				4	$9f_c$	
				3	$11f_c$	

(handwritten annotation: filter of this order will require this sampling rate)

195

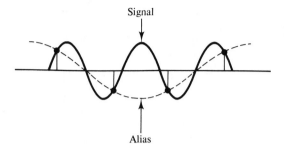

Signal

Alias

Figure 8-3 Signal Aliasing

translation of undersampled frequency components above f_0 by heterodyning with f_s. This error can be suppressed by increasing either sample rate f_s or filter rolloff rate, each contributing to its minimization. In the interest of avoiding an excessive sample rate, however, it is generally more efficient to utilize a sharper rolloff filter. This efficiency is optimized for filters in the three- to five-pole range as indicated in Table 8-1, which indicates that a minimum practical f_s/BW ratio of about ten is necessary to minimize aliasing with filter passband signal spectral occupancy as presented in Table 4-9. In the following section, however, it is shown that antialiasing sample-rate considerations are usually superseded by resolution requirements.

Residual noise at f_0 can also result in noise aliasing in the manner of signal aliasing. Ideally, -40 dB provided by an antialiasing filter at f_0 is required in sampled-data systems to minimize the transmission and conversion of noise existing at frequencies above those of the signal and to prevent broadening of the foldover spectrums at f_0 due to aliasible noise. Odd-order filters are represented in Table 8-1 because of the minimum number of operational amplifiers required in their mechanization with unity-gain lowpass networks. First- and second-order Butterworth types are included primarily as approximations to the RC filters utilized with flying-capacitor multiplexer input systems, and Besselworth filters for accurate harmonic signal-conditioning applications.

Integrating converters are coordinated with dc and sinusoidal signals because their averaging characteristic, with its attendent noise rejection, is ideal for these signals, which are frequently encountered in low-data-rate and noisy environments. This averaging characteristic is unacceptable for complex harmonic signals because of the distortion it provides to the higher-frequency components. The faster voltage-comparison converters are acceptable for all signal types but do not offer the noise-rejection capabilities of the integrating converters. However, this noise-rejection capability is available only in direct-conversion applications that do not include a sample-hold device.

$$\text{RZ } \sin x/x = \int_{-\infty}^{\infty} p(t) \cdot e^{-j\omega t} \cdot dt \qquad (8\text{-}1)$$

$$= \frac{1}{T} \int_{-\tau/2}^{\tau/2} 1 \cdot e^{-j2\pi ft} \cdot dt$$

$$= \frac{\tau}{T} \cdot \frac{\sin \pi f\tau}{\pi f\tau}$$

$$\text{NRZ } \sin x/x = \int_{-\infty}^{\infty} (1 - e^{-T}) \cdot dt \qquad (8\text{-}2)$$

$$= \frac{1}{j2\pi f} \cdot (1 - e^{-j2\pi fT})$$

$$= T \cdot \frac{\sin \pi fT}{\pi fT}$$

$$X'_{(t)} = X(t) \cdot p(t) \qquad (8\text{-}3)$$

$$= X(t) \cdot \sum_{-\infty}^{\infty} \frac{\sin x}{x} \cdot e^{j2\pi ft}$$

$$X'_{(f)} = \sum_{-\infty}^{\infty} \frac{\sin x}{x} \cdot X(f - f_s) \qquad (8\text{-}4)$$

$$\text{average RZ } \sin x/x \text{ error} = \frac{1}{2}\left(1 - \frac{\sin \pi\, BW\, \tau}{\pi\, BW\, \tau}\right) \cdot 100\% \qquad (8\text{-}5)$$

$$\text{average NRZ } \sin x/x \text{ error} = \frac{1}{2}\left(1 - \frac{\sin \pi\, BW/f_s}{\pi\, BW/f_s}\right) \cdot 100\% \qquad (8\text{-}6)$$

8-2 SAMPLE RATE AND RESOLUTION

This section shows that the sample rate for a data-conversion system is selected primarily to achieve the output accuracy of interest, and that this accuracy is usually determined by amplitude-resolution error. *Resolution* is a measure of the smallest amplitude value as a percent of full scale to which a quantity can be determined. In a data-acquisition system the resolution of a measurement is ultimately determined by the granularity of the signal due to additive noise and interference. However, in a data-conversion system it is principally determined by the conversion period T, or its inverse the sample rate f_s, as indicated by the nomograph for binary resolution of Figure 8-5. Consequently, achievable amplitude resolution is affected both by the signal-quality considerations presented in Sections 6-2 through 6-4 and by the sampling considerations relative to signal variations that follow, assuming a sufficient A/D converter output wordlength in bits for the latter.

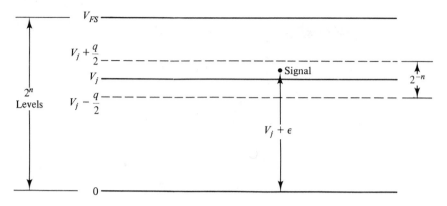

Figure 8-4 Uniform A/D Quantizing

$$Eqe = \left(\frac{1}{q} \int_{-q/2}^{q/2} \epsilon^2 \cdot d\epsilon \right)^{1/2} \qquad (8\text{-}7)$$

$$= \frac{q}{2\sqrt{3}} \quad \text{rms volts}$$

$$\text{SER}_{dB} = 10 \log \left(\frac{V_{FS}/2\sqrt{2}}{Eqe} \right)^2 \qquad (8\text{-}8)$$

$$= 10 \log \left(\frac{2^n \cdot q/2\sqrt{2}}{q/2\sqrt{3}} \right)^2$$

$$= \underset{\text{(signal)}}{6.02n} + \underset{(q \text{ noise})}{1.78}$$

Quantization of a sampled analog waveform involves the assignment of a finite number of amplitude levels corresponding to discrete values of voltage increasing from zero to some full-scale value V_{FS}, such as 10 V. The quantization interval 2^{-n} represents the least significant bit (LSB) or resolution limit for an A/D converter of n bits, with 2^n total intervals between 0 and V_{FS} of spacing $q = V_{FS}/(2^n - 1)$. Figure 8-4 describes the prevailing uniform quantizing algorithm whereby if an input signal is within the V_jth level range of $\pm q/2$, where each level has a spacing of q volts, the V_jth level is taken as the value to be encoded with a quantization error of ϵ volts. This error, which can range up to $\pm q/2$ volts, is irreducible noise added to the converted signal equivalent to the quantization-error *(Eqe)* value calculated by equation (8-7). Quantization noise has a normal probability density function centered on the quantization levels and is similar to Gaussian noise. This noise is represented in the data-conversion-system error budget as the $\pm \frac{1}{2}$ LSB A/D converter quantization error and is confined to values between $\pm q/2$.

TABLE 8-2
DECIMAL EQUIVALENTS OF 2^n AND 2^{-n}

BITS, n	LEVELS, 2^n	LSB WEIGHT, 2^{-n}	OUTPUT SER, dB
1	2	0.5	8
2	4	0.25	14
3	8	0.125	20
4	16	0.0625	26
5	32	0.03125	32
6	64	0.015625	38
7	128	0.0078125	44
8	256	0.00390625	50
9	512	0.001953125	56
10	1,024	0.0009765625	62
11	2,048	0.00048828125	68
12	4,096	0.000244140625	74
13	8,192	0.0001220703125	80
14	16,384	0.00006103515625	86
15	32,768	0.000030517578125	92
16	65,536	0.0000152587890625	98
17	131,072	0.00000762939453125	104
18	262,144	0.000003814697265625	110
19	524,288	0.0000019073486328125	116
20	1,048,576	0.00000095367431640625	122

The signal-to-error ratio (SER) of equation (8-8) defines the maximum achievable output-signal quality in power dB for an A/D converter with a noiseless input signal. This rms-signal-power-to-rms-noise-power ratio assumes sinusoidal signals and is 1.78 dB greater than the dynamic range of a data converter for the entries in Table 8-2.

A technique useful in the design and implementation of data-acquisition and conversion systems is to develop an error budget based upon the individual component and system contributions. This permits examination of system sensitivities and points where tradeoffs are warranted, as illustrated by the examples of Section 8-3. For example, an error budget shows where the largest errors are in a specific system and what effect is produced on the rms total of these errors by reducing one error component in relation to the others. A dominant error in data-conversion systems is due to the step-interpolator representation of the signal after sampling. The step interpolator is the way data are handled internally in digital processors, whereby the present sample is current data until a new sample is acquired. Figure 8-6 and equation (8-9) illustrate that a peak-to-peak sinusoidal input signal will acquire a maximum amplitude-resolution error of ΔV volts during a conversion period T. Equation (8-10) specifies the longest permissible time to achieve a specified resolution,

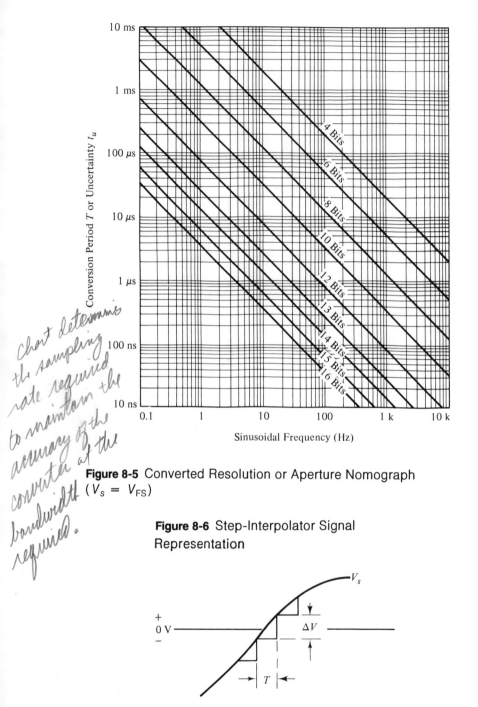

Figure 8-5 Converted Resolution or Aperture Nomograph ($V_s = V_{FS}$)

Figure 8-6 Step-Interpolator Signal Representation

Handwritten margin note: Chart determines the sampling rate required to maintain the accuracy of the converter at the bandwidth required.

where ΔV is represented by the minimum resolvable LSB increment 2^{-n} • V_{FS} from Table 8-2. For example, this specific solution shows conversion must be repeated every 99.5 μs for five-bit resolution with a 100-Hz signal, which is equivalent to a sample rate of 100 times signal BW ($f_s = 1/99.5$ μs) with $V_s = V_{FS}$. Figure 8-5 is a nomograph of equation (8-10) for signals equal to V_{FS}. Expressing the signal amplitude V_s relative to V_{FS} permits accommodation of signal variations and their influence on converted resolution.

sampling rate

to be hi enough

$$\Delta V = T \; \frac{dV_s}{dt} \tag{8-9}$$

resample signal $\quad = T \dfrac{d}{dt} \dfrac{1}{2} V_s \sin 2\pi BWt\big|_{t=0}$

fore signal change $\quad = T\pi BWV_s$

x% has occurred. $\quad T = \dfrac{2^{-n} V_{FS}}{\pi BWV_s} \tag{8-10}$

is determined by $\quad = \dfrac{(0.03125)\,(1.0)}{(3.14)\,(100\ \text{Hz})\,(1.0)}$

LSB weight. $\quad = 99.5\ \mu\text{s} \ (100\ \text{Hz to five bits})$

Sampling criteria for communication systems are essentially concerned with the ability to accurately sample and reconstruct the frequency information of signals. In contrast, it should be appreciated that measurement signals require accountable resolution of the sampled waveform and ultimately are recovered by an output-reconstruction process for display and recording purposes or a final actuator element. The relationship between sample rate and amplitude resolution is one of the fundamental tradeoffs in a sampled-data system, and it is directly affected by the interpolation method used in the reconstruction process. The developments of Chapter 9 address signal-recontruction, yielding the result that practical techniques typically offer two to three bits increase in output resolution over step-interpolator-represented data. As a result, the overall system input-to-output accuracy is determined by substituting the lower amplitude-resolution error of the reconstructed output signal for that generated by the sampled-data step interpolator. After estimating the spectrum and presampling-conditioning requirements of an input signal, therefore, we can proceed to determine what total error is allowable in the recovered output signal. Then we choose a sample rate f_s that is coordinated with the other system and component error contributions to achieve the input-to-output accuracy of interest, including consideration of the amplitude-resolution improvement realized through reconstruction.

A complex harmonic signal spectrum has a typical first-order rolloff as illustrated in Figure 1-2. Consequently, amplitude-resolution error provided by equation (8-9), which is referenced to the full-scale signal amplitude, must be scaled accordingly. If the amplitude V_s of the signal fundamental frequency is scaled to V_{FS}, then the tenth-harmonic amplitude V_s typically will be at one-tenth of V_{FS}. With reference to the complex harmonic signal bandwidth requirements of Table 8-1, therefore, the step-interpolator-represented converted resolution found by equation (8-10) has the same value at the 0.1 V_{FS} tenth harmonic as the V_{FS} fundamental for a specific sample rate. This occurs because of compensation of the increasing signal rate-of-change beyond the fundamental frequency by decreasing amplitude. Thus, adequate sampling requirements imposed at the V_{FS} fundamental frequency determine the amplitude resolution for the fundamental and its harmonics with first-order signal rolloff. The amplitude characteristics of other signal spectrums can also be worked into this development to determine the sample rate required for their amplitude resolution of interest.

Amplitude-aperture error represents the finite bound within which the amplitude of a sampled signal is acquired. Unlike amplitude resolution, which is a function of the sample rate f_s or conversion period T provided, aperture error is a function of the conversion-system aperture uncertainty time t_u and cannot be improved by output signal-reconstruction methods. The equivalent binary resolution obtained from aperture uncertainty t_u is also determined with the aid of equation (8-10) and Figure 8-5 by substituting t_u for T, and it shares sensitivity to signal amplitude and frequency with resolution. Direct-A/D conversion applications result in an aperture t_u equal to the A/D converter conversion time, which can produce a significant aperture error. Consequently, data conversion without the use of a sample-hold is best achieved by a converter mechanization that minimizes this error, such as the tracking converter of Figures 7-20 and 8-12. The use of a sample-hold device permits substitution of its substantially shorter t_u for the A/D conversion time, resulting usually in a nominal aperture error. Conversion period T is typically greater than uncertainty t_u, however, because T $(1/f_s)$ includes both A/D conversion time and a wait proportional to the number of channels multiplexed; but when equal aperture supersedes amplitude resolution.

In summary, the minimum sampling requirements tabulated in Table 8-1 that minimize aliasing in sampled-data systems are usually inadequate to represent the input signal in sampled form to the amplitude resolution of interest. Sampled analog measurement signals are also almost always recovered, to complete their destination following digital processing by means of an output reconstruction process whose interpolation method usually increases the amplitude resolution above its sampled-data step-interpoltor representation. And the input-to-output accuracy of the converted and reconstructed signal depends upon the rms of the total system and component errors, which include this improved resolution value. Consequently, the sample rate is selected to achieve the output accuracy of interest, which is typically determined by resolution.

For wide-bandwidth data such as video signals sample rate may be especially conserved as a result of output resolution improvement. For example, 256 × 256 adjacent picture elements (pixels) scanned once per second require a signal bandwidth of 131 kHz from Table 1-2 (2/event width). A minimum conversion-system throughput rate of 1.31 MHz would therefore be necessary to insure freedom from signal aliasing; it would provide approximately two-bit step-interpolator resolution, considering an f_s/BW ratio of 10 allowed by the use of a four-pole Besselworth antialiasing filter from Table 8-1. Using an identical filter for reconstruction at the same throughput rate increases the output amplitude resolution to five bits (32 brightness levels) by the methods of Tables 9-1 and 9-2, with minor effect on the spatial resolution of the picture due to filter time delay. No concessions to video source statistics are assumed in this result.

8-3 CONVERSION-SYSTEM APPLICATIONS AND ERROR ANALYSIS

The total rms system error permissible for the combined signal-conditioning and data-conversion components cannot exceed the A/D converter amplitude error values of Table 8-3 for the corresponding binary output accuracies shown to be realized. Since a data converter cannot distinguish an analog difference less than 1 LSB, its output at any point may be in error by as much as $\pm \frac{1}{2}$ LSB. Thus the A/D converter also contributes a component error requiring an output wordlength of two bits greater than the accuracy of interest typically in order to minimize its influence on the total system error budget. Table 8-4 shows the various data-conversion system budgets.

Practical considerations require that the full signal excursion range be carefully evaluated in scaling the signal-conditioning circuit gains so that the peak signal amplitude equals the A/D converter V_{FS} value. For example, a 2-volt maximum signal applied to a 10-volt V_{FS} 12-bit converter results in an awkward output whereby valid data exist only between ten bits of the converter. Consequently, a 12-bit converter may be scaled for ten bits but be capable only of eight-bit resolution with six-bit accuracy, depending upon the amplitude scaling, sample rate, signal frequency, and system error budget. Analog data-acquisition and conversion systems normally can accommodate a range of transducer input signals whose full-scale values may exist between 10 mV and 10 V, where this range is primarily determined by the maximum available amplification, which is usually 1000. It is important to note that the minimum input amplitude that can be accurately resolved is determined by $V_{error_{RTI}}$ at the amplifier input, which defines instrumentation-amplifier uncertainty.

Examination of the gamut of commercially available A/D converters reveals two general design methods, integration and voltage comparison, which

Not A/D error *see 203* *8.3*

½ LSB

TABLE 8-3
A/D CONVERTER PARAMETERS

BINARY OUTPUT (BITS)	AMPLITUDE ERROR (1 LSB) (%FS)	LSB AMPLITUDE (10 V_{FS}) (mV)
1	50.0	5000
2	25.0	2500
3	12.5	1250
4	6.2	625.0 250
5	3.1	312.6
6	1.6	156.3
7	0.8	78.2
8	0.4	39.1
9	0.2	19.6
10	0.1	9.8
11	0.05	4.9
12	0.024	2.4
13	0.012	1.2
14	0.006	0.6
15	0.003	0.3
16	0.0016	0.15
17	0.0008	0.08
18	0.0004	0.04
19	0.0002	0.02
20	0.0001	0.01

in order to realize 'x' Bit accuracy the system error specs cannot exceed the amount shown under amplitude error.

divide the conversion time range extending from 0.1 second to about 10 nanoseconds into two segments at 1 millisecond with essentially no overlap. Table 8-1 coordinates A/D converters with input signals and antialiasing requirements. From this, data-conversion systems and their input-signal-conditioning circuits can be specified and organized into analog input systems. This organization is frequently either that of a 10-μs-cycle-time (100-kHz throughput rate) multiplexed successive-approximation data-conversion system according to Figure 8-7, or a 25-ms-cycle-time (40-Hz throughput rate) flying-capacitor multiplexed dual-slope converter system as suggested by Figure 8-8. The slower conversion speeds and normal-mode noise-rejection

TABLE 8-4
DATA-CONVERSION-SYSTEM BUDGETS

BUDGET	PRIMARILY AFFECTS
Error	Output binary accuracy
Timing	Throughput rate and resolution
Cost	Component selection and tradeoffs that minimize system error

V_{diff} = 70 mV rms

V_{cm} = 1 V rms

BW = dc – 1 kHz

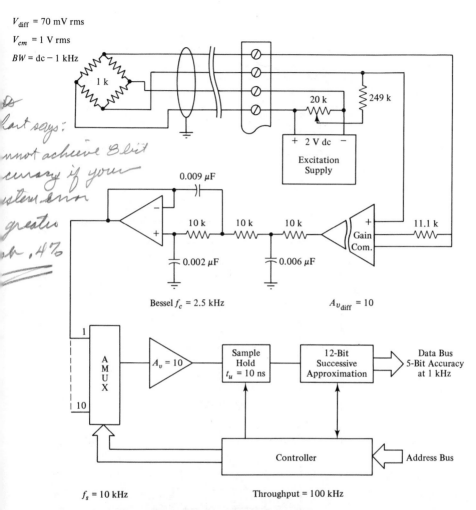

handwritten margin note: Dart says: annot achieve 8 bit curacy if your system error greater th .4%

Bessel f_c = 2.5 kHz Av_{diff} = 10

f_s = 10 kHz   Throughput = 100 kHz

Figure 8-7 High-Data-Rate Analog Input System

capability associated with integrating converters, plus the consistent accuracy provided by their self-calibrating feature, makes them ideally suited for low-data-rate dc and sinusoidal signals frequently encountered in noisy environments such as process-control applications. The faster conversion speeds offered by successive-approximation converters are better suited to the generally wider bandwidth requirements of complex harmonic signals, such as associated with biomedical and many laboratory data-acquisition applications.

The examples that follow bring together the developments of the preceding chapters for determining the component and signal error contributions to the error budget of an analog input system. These examples are

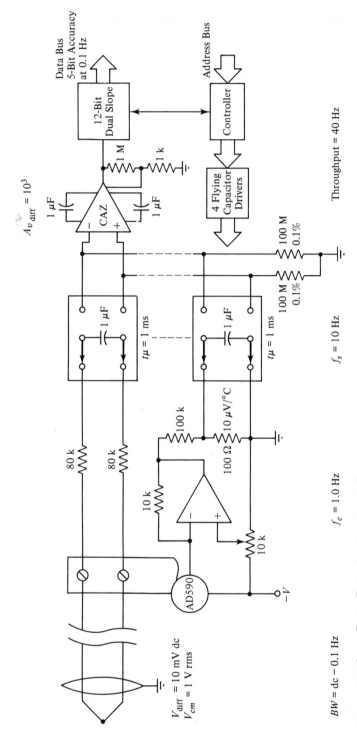

Figure 8-8 Low-Data-Rate Analog Input System

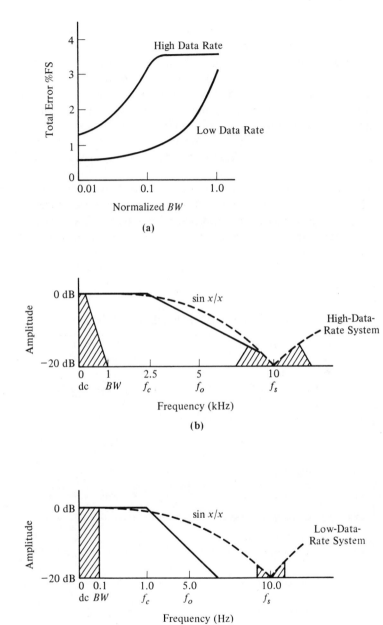

Figure 8-9 Analog Input Systems: (a) Total Error, (b) High-Data-Rate, and (c) Low-Data-Rate Frequency Spectrums

chosen to be representative of achievable performance with contemporary components for the two general signal classifications of Table 8-1 under low-level-signal measurement conditions, but without resort to complex or costly mechanizations such as specially aligned components to achieve the lowest possible errors. Users can choose from among three approaches to implement an analog input system: (1) buy devices and IC's and construct a system, (2) acquire modular/PC board systems and configure a system, or (3) purchase a complete system from a systems house. Advantage normally lies with the second approach of purchasing an engineered data-conversion-system board compatible with the specific computer with which it must interface from suppliers such as Datel-Intersil, Burr Brown, or Data Translation. Custom signal-conditioning features may then be added to the input of these conversion systems as required to achieve the performance of interest. Two complete analog input systems and their respective error budgets are shown in Figures 8-7 and 8-8 and Table 8-5.

A 0.1-Hz signal bandwidth is provided for the low-data-rate type-S

TABLE 8-5
ANALOG-INPUT-SYSTEM ERRORS AT FULL BANDWIDTH

	LOW-DATA-RATE SYSTEM DC and Sinusoidal Signals		HIGH-DATA-RATE SYSTEM Complex Harmonic Signals	
Component	Error (%FS)	Comment	Error (%FS)	Comment
Transducer	0.25	DC–0.1 Hz, V_{diff} = 10 mV dc	1.0	DC–1 kHz, V_{diff} = 70 mV rms
Interface	0.1	Junction compensation	0.01	Bridge excitation
Filter	0.25	1-Pole RC, 1.0 Hz f_c	0.75	3-Pole Bessel, 2.5 kHz f_c
Amplifier	0.051	CAZ, $A_{v_{\text{diff}}} = 10^3$	0.086	Isolation, $A_{v_{\text{diff}}} = 10$
Signal Quality	0.63	60-Hz interference	0.2	DC, 60-Hz, random noise
Aperture	0.03	Relay bounce dependent	0.0003	Sample-hold uncertainty
Resolution	3.1	$\dfrac{f_s}{BW} = 100$, V_{FS} at BW	3.1	$\dfrac{f_s}{BW} = 10$,0.1 V_{FS} at BW
Aliasing	0.01	Translated residual 60 Hz	0.06	From noise at f_0
Multiplexer	0.01	Flying capacitor	0.01	CMOS switches
Sample Hold	0.49	Acquisition dependent	0.02	Acquisition dependent
Sin x/x	0.009	Average attenuation	0.82	Average attenuation
A/D	0.06	12-Bit integrating	0.07	12-Bit successive approximation
RMS Total	3.22%	5-Bit accuracy	3.44%	5-Bit accuracy

thermocouple transducer and 1-kHz for a complex harmonic high-data-rate signal with a maximum V_{FS} frequency of 100 Hz. Antialiasing filters precede the multiplexers for both systems, and a minimum five-bit accuracy is sought for each signal. The instrumentation amplifiers have the gains shown in Figures 8-7 and 8-8, and 10^5 CMRR values (provided by the flying capacitors in Figure 8-8). Also, amplifier prefiltering reduces overload vulnerability in Figure 8-8, whereby distributing the gain between the amplifiers in Figure 8-7 reduces overload problems with out-of-band signals. Amplifier error as a percent of full-scale output is seen to depend upon the value of $A_{v_{diff}}$ from equation (3-19). For the signal-quality calculations common-mode interference is dc plus possible 60 Hz and random noise in the high-data-rate system, and essentially 60 Hz in the low-data-rate system. The filters in these examples primarily guard against alias causing out-of-band interference rather than providing SNR improvement. Transducer-loop noise $V_{N_{p-p}}$ is dominated by amplifier V_n in the low-data-rate system, and contact noise V_c in the high-data-rate system, referring to Section 3-2. The signal-quality efficiency-factor K multiplier for the high-data-rate channel is 0.9 for all amplifier output SNR values above 100 from Figure 6-6, but SNR is not improved beyond that at the amplifier output.

The RC antialiasing filter preceding the flying-capacitor multiplexer provides an example whereby the value of f_c to provide adequate attenuation at the folding frequency f_0 competes with the necessity of allowing a sufficient number of time constants for signal settling between samples. Coordination with the parameters from Table 4-9 indicates the choice of a one-pole RC filter with a cutoff frequency of 1 Hz. This minimizes filter gain error in the signal passband, provides acceptable signal settling during each acquisition period, and satisfies practical antialiasing requirements. The filter RC values are determined from equation (6-11), and the average passband gain error is 0.25%. A 1-Hz f_c and 10-Hz f_s allow about half the 160-ms RC time constant during any 100-ms channel conversion period (10 Hz) for signal settling to within 63% of the amplitude change during the 75-ms acquisition periods (100 ms minus 25 ms conversion-cycle time) by Figure 2-18. By equation (8-9) the signal-amplitude change ΔV during any 25-ms conversion-cycle time is 78 mV for a 10-V_{FS} signal at 0.1 Hz. The settling error is therefore $\Delta V63\%$ or 49 mV (0.49% FS) at the end of the 75-ms acquisition period and is equivalent to 1 LSB of an eight-bit A/D conversion (0.4%) by Table 8-3. This acquisition error is properly included in the sample-hold error budget.

The 10-Hz sample rate allows four identical multiplexed channels for the low-data-rate system 40-Hz throughput rate. Direct conversion by the integrating dual-slope A/D converter provides useful noise rejection to the small residual output chopper noise of the CAZ amplifier. However, the flying-capacitor multiplexer sample-hold function prevents realization of this noise-rejection capability for the input signal. The aperture uncertainty time of this sample-hold is also equal to the relay bounce time, which is typically in the 1-ms range. Either equation (8-10) or extrapolation of Figure 8-5 yields five-bit resolution at the V_{FS} amplitude 0.1-Hz signal bandwidth with a sample rate of 100 times this bandwidth $[T = 1/(10 \text{ Hz})]$.

Interestingly, three $\sin x/x$ functions are generated in the low-data-rate conversion system by flying-capacitor sampling, the CAZ amplifier, and integrating A/D converter. However, only the sampling function is of consequence. Provisions of both systems result in negligible signal-aliasing errors and consequently involve only noise-aliasing considerations. An aliasing error of 0.01% is obtained from the 0.63% residual 60 Hz coherent interference following filtering and downward frequency translation by f_s to reach the signal passband. This heterodyne operation is defined by successive half-angle trigonometric expansions between the interference and sampling frequencies. Aperture error is 0.03% for the low-data-rate system and a negligible 0.0003% for the complex harmonic signal of the high-data-rate system. The transducer-loop noise $V_{N_{p-p}}$ of $43\mu V$ in the high-data-rate system is dominated by the contact noise of equation (3-13), resulting from a 1-mA dc current from the 1-V dc V_{cm} transducer excitation impressed across the 1-k R_s. Internal noise for both systems is determined at the more severe conditions existing at 10 Hz discussed in Section 3-2.

The 0.2% residual interference at f_0, attenuated by -10 dB from the filter and $\sin x/x$ function, provides a simplified worst-case expression for aliasing error of 0.06% with uniformly distributed random noise in the high-data-rate system. A 10-kHz $[T = 1/(10 \text{ kHz})]$ sample rate for the high-data-rate system also provides five-bit resolution at its 1-kHz bandwidth with a 0.1-V_{FS} signal amplitude, and can multiplex ten inputs with its 100-kHz throughput rate. The interface circuit for the high-data-rate system is dominated by a 0.01% regulated transducer power supply. The Analog Devices AD590 temperature sensor is capable of providing cold-junction compensation with an accuracy of 0.1% of 1000°C full scale for an entire terminating strip of thermocouples in the low-data-rate system. By appropriately scaling and offsetting its 1-$\mu A/°K$ output, a 10-$\mu V/°C$ type-S thermocouple compensation signal is provided which is sampled as another input signal. Both systems provide effective isolation against external fault voltages. The signal-conditioning overload detector described by Figure 6-12 may also be added as an option to indicate signal-amplitude limiting.

These analog-input-system error budgets, summarized in Table 8-5, highlight sensitivities and point where useful tradeoffs can be made to reduce the total rms error for each system. The low-data-rate system has an accuracy of five bits for dc and sinusoidal signals at 0.1 Hz, and the high-data-rate system five-bit accuracy for complex harmonic signals at 1 kHz. These accuracies are primarily attributable to the amplitude-resolution errors provided by the step-interpolator representation of the sampled data at sample rates of 100 times the highest V_{FS} frequencies. Other contributions are seen to be due mainly to transducer non-linearity, acquisition, filter component error, signal quality, and $\sin x/x$ attenuation. In practice the eight MSB data lines from the 12-bit A/D converters may be used, such as with eight-bit processors, with essentially no change in these results because the four LSB's typically represent less than the total residual error of either system described in Figure 8-9.

It should be appreciated that these error analyses were calculated at the worst-case highest signal frequency defining the required bandwidth, and that output accuracy improves as signal frequency decreases with the same f_s. For example, at one-tenth bandwidth the low-data-rate system error improves to approximately 0.79% or seven-bit accuracy, essentially due to improvement in amplitude resolution and acquisition errors. However, at one-tenth bandwidth the high-data-rate system error is essentially unchanged at 3.34% because of the constant amplitude resolution with first-order increasing signal harmonic amplitudes to the V_{FS} frequency of 100 Hz with f_s constant. This error relationship is illustrated for both systems between their 1% and 100%-bandwidth points in Figure 8-9, describing an asymptotic relationship largely attributable to decrease of amplitude-resolution error down to the residual system errors as the signal bandwidth is reduced. The 1% BW errors for the high- and low-data-rate systems are 1.28% and 0.73%, respectively. Additional error terms may also be added, depending upon their availability and significance to specific tasks, such as A/D converter harmonic and intermodulation distortion errors in spectrum analyzer applications.

LOW-DATA-RATE SYSTEM	HIGH-DATA-RATE SYSTEM
Transducers	
Type-S thermocouple with 0.25%FS error, V_s = 10 mV dc at $1000°$C to 0.1 Hz BW, V_{cm} = 1 V rms 60-Hz interference, R_s = 100Ω.	Piezo-resistor strain-gage sensor, complex harmonic V_s = 70 mV rms dc to 100 Hz, −20 db rolloff to V_s = 7 mV rms at 1-kHz BW, 1.0%FS error and R_s = 1 k. DC, 60 Hz, and random V_{cm} = 1 V rms.
Interfaces	
One-pole RC filter, 10 μV/°C cold-junction compensation with 0.1% error.	Two-volt dc transducer excitation regulated to 0.01%.
Filters	
One-pole RC filter with 0.25% average amplitude error to 0.1 Hz, f_c = 1.0 Hz, for dc and sinusoidal signals (Table 4-9).	Three-pole Bessel active lowpass filter, 0.75% amplitude error with complex harmonic signals to 1 kHz, f_c = 2.5 kHz (Table 4-9).

LOW-DATA-RATE SYSTEM	*HIGH-DATA-RATE SYSTEM*
	Amplifiers	

Commutating autozero amplifier:		Isolation amplifier:	
$dT = 20°C$		$dT = 20°C$	
$f_{hi} = 10$ Hz		$f_{hi} = 3.3$ kHz	
$dV_{os}/dT = 0.01\ \mu V/°C$		$dV_{os}/dT = 2\ \mu V/°C$	
$I_{os} = 150$ pA		$I_{os} = 50$ pA	
$\dfrac{dI_{os}}{dT} = 1$ pA/°C		$\dfrac{dI_{os}}{dT} = 0.1$ nA/°C	
$6.6\ V_n \sqrt{10\text{ Hz}} = 4\mu V_{p\text{-}p}$		$6.6\ V_n \sqrt{3.3\text{ kHz}} = 15\ \mu V_{p\text{-}p}$	
$f(A_v) = 10$ ppm		$f(A_v) = 500$ ppm	
$dA_v/dT = 15$ ppm/°C		$dA_v/dT = 35$ ppm/°C	
CMRR $= 10^5$ incircuit		CMRR $= 10^5$ incircuit	
$A_{v_{\text{diff}}} = 10^3$		$A_{v_{\text{diff}}} = 10$	
V_{FS} output $= 10$ V		V_{FS} output $= 10$ V	
$R_s = 100\Omega$		$R_s = 1$ k	
$V_{N_{p\text{-}p}} = 6.6\ V_n \sqrt{10\text{ Hz}}$		$V_{N_{p\text{-}p}} = 6.6\left[(10^{-6}\sqrt{1/10\text{ Hz}}\ 1\text{k}\ 1\text{mA})^2 + V_n^2\right]^{\frac{1}{2}}\sqrt{3.3\text{ kHz}}$	

$\dfrac{dV_{os}}{dT}\cdot dT$	$0.2\ \mu V$	$\dfrac{dV_{os}}{dT}\cdot dT$	$40\ \mu V$
$I_{os}\cdot R_s$	$0.015\ \mu V$	$I_{os}\cdot R_s$	$0.05\ \mu V$
$\dfrac{dI_{os}}{dT}\cdot dT\cdot R_s$	$0.002\ \mu V$	$\dfrac{dI_{os}}{dT}\cdot dT\cdot R_s$	$2.0\ \mu V$
$V_{N_{p\text{-}p}}$	$4.0\ \mu V$	$V_{N_{p\text{-}p}}$	$43\ \mu V$
$f(A_v)\cdot \dfrac{V_{FS_{\text{output}}}}{A_{v_{\text{diff}}}}$	$0.1\ \mu V$	$f(A_v)\cdot \dfrac{V_{FS_{\text{output}}}}{A_{v_{\text{diff}}}}$	$500\ \mu V$
$\dfrac{dA_v}{dT}\cdot dT\cdot \dfrac{V_{FS_{\text{output}}}}{A_{v_{\text{diff}}}}$	$3.0\ \mu V$	$\dfrac{dA_v}{dT}\cdot dT\cdot \dfrac{V_{FS_{\text{output}}}}{A_{v_{\text{diff}}}}$	$700\ \mu V$
RMS $V_{error_{RTI}}$	$5.1\ \mu V$	RMS $V_{error_{RTI}}$	$861\ \mu V$
$\epsilon_{ampl_{\%FS}}$	0.051%	$\epsilon_{ampl_{\%FS}}$	0.086%

LOW-DATA-RATE SYSTEM	HIGH-DATA-RATE SYSTEM

Signal Quality

input SNR $= \left(\dfrac{V_{diff}}{V_{cm}}\right)^2$

$= \left(\dfrac{10^{-2}v}{1v}\right)^2 = 10^{-4}$ (6-3)

input SNR $= \left(\dfrac{V_{diff}}{V_{cm}}\right)^2$

$= \left(\dfrac{7 \times 10^{-3}v}{1v}\right)^2 = 4.9 \times 10^{-5}$ (6-3)

filter $SNR_{60\ Hz}$

$= \dfrac{\text{input SNR}}{(\text{filter attn.})^2} = \dfrac{10^{-4}}{(0.015)^2}$ (6-6)

$= 0.5$ (200% @ 1 σ by equation 6-1)

amplifier SNR

$= (\text{input SNR})(\text{CMRR})^2$ (6-4)

$= (4.9 \times 10^{-5})(10^5)^2$

$= 4.9 \times 10^5$ (0.2% @ 1 σ Table 6-2)

flying capacitor SNR

$= (\text{filter SNR})(\text{CMRR})^2$ (6-4)

$= (0.5)(10^5)^2$

$= 5 \times 10^4$ (0.63% @ 1 σ)

$\epsilon_{signal_{\%FS}} = 0.63\%$ (6-7)

filter SNR_{random}

$= \text{amplifier SNR} \cdot K \cdot \dfrac{\text{amplifier } f_{hi}}{\text{filter } f_c}$

$= (4.9 \times 10^5)(0.9) \dfrac{3.3\ kHz}{2.5\ kHz}$ (6-5)

$= 6.15 \times 10^5$ (0.2% @ 1σ)

filter $SNR_{60\ Hz}$

$= \dfrac{\text{amplifier SNR}}{(\text{filter attn.})^2} = \dfrac{4.9 \times 10^5}{(1)^2}$ (6-6)

$= 4.9 \times 10^5$ (0.2% @ 1 σ)

$\epsilon_{signal_{\%FS}} = 0.2\%$ (6-7)

Aperture

$2^{-n} = \dfrac{\pi BW t_u\ V_s}{V_{FS}}$ (8-10)

$= \dfrac{(3.14)(0.1\ Hz)(10^{-3}\ s)(10\ mV)}{10\ mV}$

$= 0.000314$

$\epsilon_{aperture_{\%FS}} = 0.03\%$

$2^{-n} = \dfrac{\pi BW t_u\ V_s}{V_{FS}}$ (8-10)

$= \dfrac{(3.14)(1\ kHz)(10^{-8}\ s)(7\ mV)}{70\ mV}$

$= 0.000003$

$\epsilon_{aperture_{\%FS}} = 0.0003\%$

LOW-DATA-RATE SYSTEM	*HIGH-DATA-RATE SYSTEM*

Resolution

$$2^{-n} = \frac{\pi BW \, V_s}{V_{FS} f_s} \quad (8\text{-}10)$$

$$= \frac{(3.14)(0.1 \text{ Hz})(10 \text{ mV})}{(10 \text{ mV})(10 \text{ Hz})}$$

$$= 0.031$$

$$\epsilon_{\text{resolution}_{\%FS}} = 3.1\%$$

$$2^{-n} = \frac{\pi BW \, V_s}{V_{FS} f_s} \quad (8\text{-}10)$$

$$= \frac{(3.14)(1 \text{ kHz})(7 \text{ mV})}{(70 \text{ mV})(10 \text{ kHz})}$$

$$= 0.031$$

$$\epsilon_{\text{resolution}_{\%FS}} = 3.1\%$$

Aliasing

Downward frequency translation of the 0.63% residual 60 Hz coherent interference by the f_s of 10 Hz to the dc–0.1 Hz signal passband provides 0.63% $(f_{\text{coherent}} \, X f_s)^6$ = 0.01% aliasing error.

For 0.2% random residual interference, aliasing error is adopted as the 0.06% amplitude at f_o following −10 dB(0.3) attenuation from filtering and the sin x/x function.

Multiplexers

Flying capacitor:		CMOS switch:	
Transfer error	0.01%	Transfer error	0.01%
Crosstalk	0.001	Crosstalk	0.001
Thermal offset	0.001	Leakage	0.001
RMS total $_{\%FS}$	0.01%	RMS total $_{\%FS}$	0.01%

Sample Holds

Acquisition error	0.49%	Acquisition error	0.01%
Droop	0.01	Droop (25 μV/μs) (6 μs hold)	0.0015
RMS total $_{\%FS}$	0.49%	Offset (50 μV/°C) (20° C)	0.014
		Dielectric absorption	0.01
Flying capacitor	t_u = 1 ms	Feedthrough	0.005
		Hold-jump error	0.001
		RMS total $_{\%FS}$	0.02%
		Aperture uncertainty	t_u = 10 ns

Sin x/x

Average NRZ sin x/x error

$$= \frac{1}{2}\left(1 - \frac{\sin \pi BW/f_s}{\pi BW/f_s}\right) \cdot 100\% \quad (8\text{-}6)$$

$$= \frac{1}{2}\left(1 - \frac{\sin \pi/100}{\pi/100}\right) \cdot 100\%$$

$$= 0.009\%$$

Average NRZ sin x/x error

$$= \frac{1}{2}\left(1 - \frac{\sin \pi BW/f_s}{\pi BW/f_s}\right) \cdot 100\% \quad (8\text{-}6)$$

$$= \frac{1}{2}\left(1 - \frac{\sin \pi/10}{\pi/10}\right) \cdot 100\%$$

$$= 0.82\%$$

A/D Converters

12-Bit dual slope:		12-Bit successive approximation:	
Quantization error $\left(\pm\frac{1}{2}\text{LSB}\right)$	0.012%	Quantization error $\left(\pm\frac{1}{2}\text{LSB}\right)$	0.012%
Differential linearity $\left(\pm\frac{1}{2}\text{LSB}\right)$	0.012	Differential linearity	
Gain tempco (25 ppm/°C)		$\left(\pm\frac{1}{2}\text{LSB}\right)$	0.012
(20°C)	0.050	Linearity tempco	
Offset tempco (2 ppm/°C)		(2 ppm/°C) (20°C)	0.004
(20°C)	0.004	Gain tempco	
Long-term change	0.025	(20 ppm/°C) (20°C)	0.040
RMS total %FS	0.06%	Offset (5 ppm/°C) (20°C)	0.010
		Long-term change	0.050
		RMS total %FS	0.07%

8-4 CONVERSION-SYSTEM COMPUTER INTERFACING

Data-conversion systems generally employ one of two computer interfacing methods. One is device-initiated I/O controlled by committed logic separate from the processor; the second is program-initiated memory-mapped I/O, which has simpler hardware requirements but reduces addressable computer memory. Basic I/O alternatives are block-diagrammed in Table 8-6. The I/O

TABLE 8-6
I/O CONTROL COMPARISONS FOR 1-μs PROCESSOR CYCLE

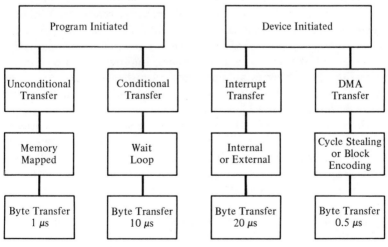

cycle times for all are strongly software-dependent. In addition, two formats are available for data transmission between peripherals such as data-conversion systems and processors—remote serial and close-in parallel. For high data rates the conversion system should be located near the processor with a parallel interface. However, if data are gathered some distance away, a serial interface perhaps using a universal asynchronous receiver/transmitter (UART) is indicated. Remote data transmission is developed in Chapter 9.

For device-initiated parallel interfacing to microcomputers, programmable peripheral interface devices are available to provide for addressing, interrupts, and data transfer under software control. Such devices as the Motorola 6821 and Intel 8255 communicate with a microprocessor over an eight-bit bidirectional data bus and provide two bidirectional eight-bit input/output ports for I/O. Figure 8-10 provides an example of a combined parallel and serial interface with a programmable interface device. The Intersil 8052A/7103A integrating monolithic A/D converter and a National 74C157 DMUX provide the necessary signals for a simple low-data-rate conversion system for both I/O modes. Software requirements include an initialization subroutine to load the programmable interface control and data-direction registers on power-up or at the user program beginning. Additional code is required to provide for data transfer, which may be either an interrupt-service routine or a data-request wait-loop data-transfer subroutine under programmed I/O control. From Table 8-6 a wait-loop conditional data transfer requires one data-conversion system cycle (10 μs) per byte.

Interrupt-driven systems are frequently employed in real-time data-conversion operations, such as encountered in process-control applications, permitting processor interrupts to modify the program sequence on an as-needed basis. Interrupts may be single-line or multilevel, the latter usually decoded by devices such as the Intel 8259 interrupt controller. Within these two general categories are maskable and nonmaskable types (for power-failure detection). These can be further subdivided into internal or external (software or hardware) initiation. The term *vectored* implies that, upon interrupt initialization, branching is automtic to a furnished service route address. *Nonvectored* interrupts halt the processor and institute a polling routine to identify which peripheral is to be serviced.

Direct memory access (DMA), or nonprocessor interrupt, transfers data directly to a computer memory location by controlling the data bus during a commandeered machine cycle. Additional hardware is required to implement this device-initiated interfacing method, such as the Motorola 6844 or Intel 8257 DMA controllers, but the time saving during data transfer can be substantial. For example, if an interrupting procedure is used, the processor can perform other tasks until a data-transfer interrupt causes the processor to branch to the interrupt-service routine and perform the transfer operation. For a 1-μs processor-instruction cycle time, a 20-instruction service routine will require 20-μs to accomplish the transfer of one byte from a programmable

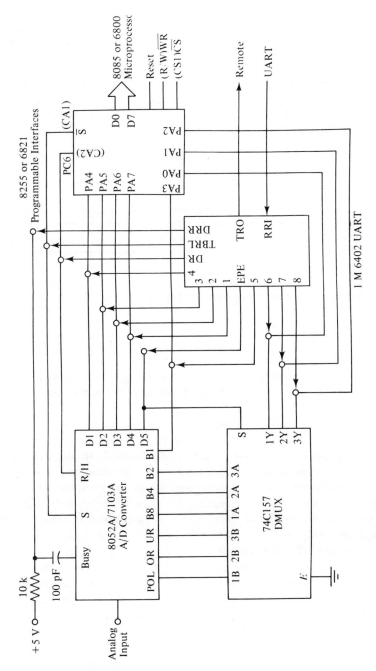

Figure 8-10 Microcomputer Parallel and Serial Analog Input

217

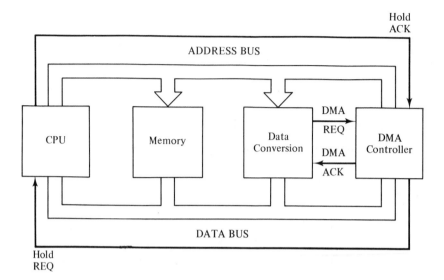

Figure 8-11 DMA-IO-Memory Interfacing

interface device. Using DMA techniques with a 500-ns memory access, this byte transfer can be reduced to 0.5 μs. The additional complexity required for the DMA interface is usually justifiable only in applications requiring high-speed data transfers. A DMA-controlled interface is diagrammed in Figure 8-11.

Program-initiated I/O usually will be slower than device-initiated I/O because the processor must enter a wait loop of some duration following a data request. This wait can extend to tens of milliseconds in the case of the conversion time of an integrating A/D converter. Operation can be structured to be unconditional, where the conversion system is always ready for communication, or conditional, where data transfer occurs only when the conversion system is ready for communication. Conditional transfer generally has an appreciable wait loop, which is inefficient. Unconditional-transfer memory-mapped I/O, described in the following two examples, offers speeds between those of the previous interrupt-initiated and DMA transfers.

Memory-mapped I/O is accomplished using individual address lines as device-enable lines. Since processor memory is organized from the lowest addresses, lines used for memory mapping normally are the most-significant-bit-address lines. This provides the user with additional instructions for dealing with I/O devices by utilizing memory-reference instructions. However, this technique does reduce addressable-processor memory in proportion to the address lines committed to I/O control. Two data-conversion system configurations are considered: (1) a parallel conversion system of multiple A/D converters that convert continuously without sample-hold circuitry and are wire-ORed to the data bus with three-state outputs, and (2) a random-addressed multiplexed system with sample-holds and on-board output memory.

The parallel conversion system shown in Figure 8-12 has simple control requirements consisting of an address decoder for enabling the selected converter. With this arrangement the latest data can be accessed as if they were in main memory. A memory-transfer instruction cycle is typically a microsecond or so, depending upon the processor. The least significant bits of the address are decoded to select the addressed channel, and the most significant address bits, which select the processor memory page, are also examined by a magnitude comparator to provide conversion system handshaking functions. In addition to enabling the address decoder, these MSBs gate a tristate inverter in coincidence with a memory-read command RD to prevent

Figure 8-12 Tracking A/D Parallel Conversion System

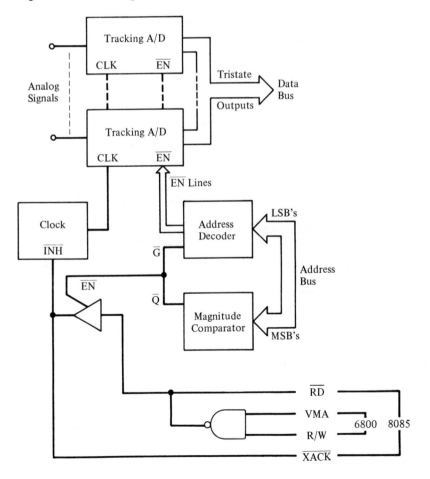

output change during data access, and to return an acknowledgment to the processor when the addresss is correct. The tracking A/D converter is well suited for this application. For example, a converter such as the Datel-Intersil ADC-856 with a 1-MHz clock can track a V_{FS} input signal up to $2^{-10} \cdot 1$ MHz for a 1-kHz conversion rate. The tracking speed that corresponds to 1 LSB is the 1 μs maximum clock rate, which in this direct-conversion application also corresponds to the aperture uncertainty t_u. Consequently, the available signal bandwidth and binary resolution are determined directly from Figure 8-5. Preconversion signal conditioning may be included as required to provide the signal quality of interest.

Figure 8-13 Random-Addressed Multiplexed Conversion System with Memory

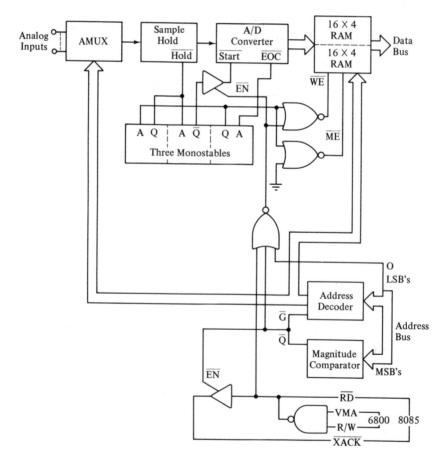

TABLE 8-7
MULTIPLEXED CONVERSION-SYSTEM TIMING

PARAMETER	MULTIPLEXER (μs)	SAMPLE HOLD (μs)	A/D (μs)
Settling	2 (overlapped)		
Acquisition		6	
Conversion			4

The random-addressed multiplexed system of Figure 8-13 offers a signal bandwidth that is dependent upon the data-conversion-system cycle time. This is typically in the range of 10 μs, as tabulated in Table 8-7; and Table 7-5 tells us that it requires a successive-approximation A/D converter. The inverse of this provides a 100-kHz system-throughput rate. This rate is apportioned among the multiplexed channels by the number of samples f_s necessary to preclude aliasing and obtain the resolution of interest. The requirement for greater resolution or signal bandwidth simply calls for higher-performance components. Complete multiplexed data-conversion systems are available with cycle times to 4 μs for a quarter-megahertz throughput rate.

Figure 8-14 diagrams a typical multiplexed system conversion cycle. The multiplexer switches channels and settles before the sample hold acquires a new sample, and the S/H is then allowed to settle before the A/D

Figure 8-14 Multiplexed Conversion System Timing Diagram

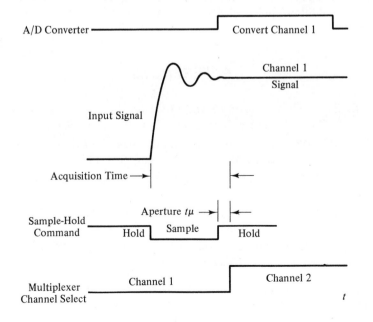

conversion is initiated by means of monostable triggering to generate the required device timing sequence. Timing jitter among the various data-conversion components is normally negligible as a source of system error. The output memory shown in Figure 8-13 permits accessing this data-conversion peripheral without requiring the processor to enter a wait mode following a data request. Consequently, instead of 100% processor dedication to the data-conversion function during this interval, the data-transfer time is only that for a memory-transfer instruction cycle. This is typically of the order of a microsecond, depending upon the processor. Latest data are always present within this memory and are updated at the rate of the conversion-system cycle time. When the memory is interrogated, the updating process is momentarily suspended while the data appear on the output data bus from the RAMs.

Measurement applications involving multiple interactive peripherals can benefit from digital interconnection employing one of the digital interfacing standards. Perhaps prominent among these is the IEEE Standard 488,* which is the same as ANSI Standard MC 1.1 and HP-IB. This standard specifies 16 signal lines grouped in eight data lines, three byte-transfer control lines, and five general interface management lines. Up to 15 peripherals or instruments may be interconnected to one bus at up to 50-foot distance with a 500-kbps data rate with one talker and up to 14 listeners at a time. Driver and receiver circuits are TTL-compatible, and the connector is specified. A modular interface standard utilized widely for nuclear instrumentation is CAMAC,† which provides both mechanical and electrical interconnection specifications. Card-edge connectors are normally used in this system with 64 lines grouped in 24 read lines, 24 write lines, 14 power and ground lines, and two strobe lines. In addition, pulse electrical parameters and dataway timing are specified.

REFERENCES

1. E. ZUCH, *Data Acquisition and Conversion Handbook*, Datel-Intersil, 11 Cabot Boulevard, Mansfield, Mass., 02048, 1979.

2. B. M. GORDON, *The Analogic Data-Conversion Systems Digest*, Analogic, Audubon Road, Wakefield, Mass., 01880, 1977.

3. J. SHERWIN, "Simplify Analog/Computer Interfacing," *Electronic Design*, August 16, 1977.

4. B. REYNOLDS, "Guide to Analog I/O Boards—Part I," *Instruments and Control Systems*, August 1978.

5. B. REYNOLDS, "Guide to Analog I/O Boards—Part II," *Instruments and Control Systems*, September 1978.

* IEEE, 345 East 47th Street, New York, N.Y. 10017.
† ERDA, Washington, D.C. 20460, publications TID-25875 through 25877.

6. M. HORDESKI, "Interfacing Microcomputers in Control Systems," *Instruments and Control Systems*, November 1978.

7. *The Application of Filters to Analog and Digital Signal Processing*, Rockland Corporation, 230 W. Nyack Road, West Nyack, N.Y. 10994, 1976.

8. *Technical Memorandum: Antialiasing Filters in Digital Data Collection Systems: A New Approach*, Electronic Instrumentation and Technology, 100 Glenn Drive, Sterling, Va. 22170.

9. *Analog-Digital Conversion Handbook*, Analog Devices, Norwood, Mass., 02062, 1972.

10. B. M. GORDON, "Digital Sampling and Recovery of Analog Signals," *Electronic Equipment Engineering*, May 1970.

11. D. F. HOESCHELE, JR., *Analog-to-Digital, Digital-to-Analog Conversion Techniques*, John Wiley, New York, 1968.

12. L. SOLOMON and E. ROSS, "Educating Dumb Data Acqusition Subsystems," *Digital Design*, November 1976.

13. D. STANTUCCI, "Data Acquisition Can Falter Unless Components Are Well Understood," *Electronics*, November 13, 1975.

14. D. STANTUCCI, "Maneuvering for Top Speed and High Accuracy in Data Acquisition," *Electronics*, November 27, 1975.

15. G. E. TOBEY, "Ease Multiplexing and A/D Conversion," *Electronic Design*, April 12, 1973.

16. N. BURSTEIN, "What to Look for in Analog Input/Output Boards," *Electronics*, January 19, 1978.

17. L. W. GARDENSHIRE, "Selecting Sample Rates," ISA Journal Reprint, April 1964.

PROBLEMS

8-1. A spectrum analyzer displays the following sampled RZ signal. Determine f_s and the width of the sampling pulse.

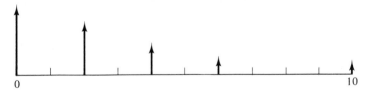

1 kHz/division

8-2. A dc digital panel meter incorporating integrating A/D conversion with an f_s of 60 Hz is used in Europe where the line frequency is 50 Hz. Determine the interference rejection at 50 Hz.

8-3. Determine the highest frequency that can be accommodated by a data-conversion system with a 40-kHz throughput rate for eight-bit resolution and a V_{FS} signal.

8-4. Evaluate the error terms of Table 8-5 for the low-data-rate system at one-tenth the 0.1-Hz signal bandwidth to verify that point on the error curve of Figure 8-9. The sample-hold acquisition error term is determined by multiplying equation (8-9) by the input-circuit RC settling time available during each signal-acquisition period defined in Section 8-3.

8-5. A video signal is to be digitally encoded to provide a per-frame resolution of 400 lines by 250 picture elements per line. For eight step-interpolator-represented brightness levels per element and ten frames per second, determine the signal bandwidth and required conversion rate for a V_{FS} signal.

8-6. A deep-space biological probe telemetry channel outputs an undefined-type, dc -1-kHz, 1-V peak analog signal, which is to be stored digitally for later analysis. Design a premium conversion system for this channel, select a sample rate that provides ten-bit amplitude resolution, specify components and their parameters, show the frequency spectrum, and calculate the conversion-system rms error budget. Utilize an isolation amplifier and component error calculations from previous chapters where applicable.

signal recovery and distribution

9-0 *INTRODUCTION*

Signal reconstruction and distribution, reversed in sequence from acquisition and conversion, substantially resemble each other in design. In this chapter mean-squared-error is developed as the criterion for quantitative evaluation of various signal-recovery methods. This error is a function of the quantization process originating in the A/D converter which is reconstituted by the analog output system DAC and minimized with presented interpolation techniques to improve output signal resolution. We then coordinate this criterion with the other system error contributions to provide a design method-of-approach for signal recovery systems.

The chapter also considers analog signal-distribution circuits, including demultiplexers, and both voltage and current signals. Instrumentation signal-transmission methods are then developed for applications when distances are such as to require them. Serial-digital line driver and receiver applications are presented for ranges to 1 mile, and both wire and radio limited-distance MODEM examples shown for ranges to 10 miles.

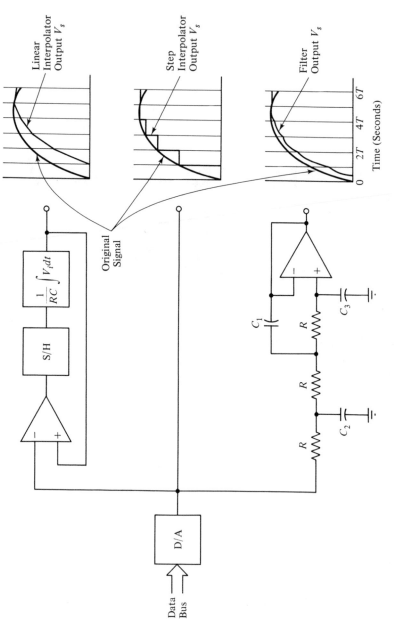

Figure 9-1 Signal-Reconstruction Methods

226

9-1 DATA-RECOVERY METHODS

The recovery of a continuous-time analog signal from a discrete-time digital signal is relevant to a number of applications. Providing output signals for actuators in process-control applications, and analog signals for display and recording purposes, are typical examples. Signal reconstruction may be analyzed from either the time-domain or frequency-domain perspective. In time-domain terms, reconstruction is similar to interpolation and extrapolation procedures in numerical analysis, with the criterion being the generation of a smooth curve that passes through the given data points. In the frequency domain, signal reconstruction may be viewed as filtering by a lowpass function to remove the periodic image spectrums, which is equivalent to averaging between data points. Important criteria are low distortion in the filter passband and high attenuation beyond, thereby eliminating possible nonlinear effects due to high-frequency components in postrecovery circuits. Figure 9-1 shows direct D/A-converter signal recovery, and signal-recovery methods employing both a linear interpolator and lowpass active filter. Note that reconstructed output signals are delayed one sample period $(1/f_s)$ in the case of the linear interpolator and signal delay is proportional to the number of filter poles and the cutoff frequency f_c with filter reconstruction.

The general sampling principles developed by Nyquist[1] require imprac-tical to realize ideal impulse sampling and ideal lowpass filtering in order to achieve perfect signal recovery—represented by Figure 9-2. An extension of this is necessary, therefore, to quantify the amplitude resolution error of real sampled data when practical signal reconstruction methods are considered. The relationship between sampling and resolution was developed in Section 8-2. The relationship between sampling rate and accuracy associated with sampled data was introduced in Chapter 8. We now extend this relationship to include output interpolation methods efficient in further recovering the step-interpolator-represented sampled data. The output of a D/A converter may be viewed as a zero-order step interpolator, as it uses polynomials of order zero for signal interpolation. The first-order linear interpolator reconstructs the output signal with increased resolution by generating a line segment with a

Figure 9-2 Signal Sampling and Reconstruction

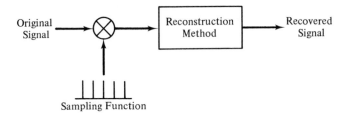

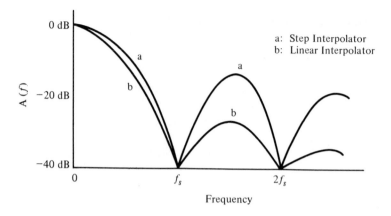

Figure 9-3 Interpolator Amplitude Responses

slope proportional to the difference between consecutive samples. It therefore interpolates between samples already received to improve output resolution at the expense of a delay of one sample period. The amplitude response of these zero-order and first-order interpolators is provided by equations (9-1) and (9-2) and plotted in Figure 9-3.

$$\text{direct-D/A } A(f) = \frac{\sin \pi f/f_s}{\pi f/f_s} \qquad (9\text{-}1)$$

$$\text{linear interpolator } A(f) = \left(\frac{\sin \pi f/f_s}{\pi f/f_s} \right)^2 \qquad (9\text{-}2)$$

Although an A/D converter accomplishes the quantization of an analog waveform, the D/A converter provides the awareness of this quantization by means of its output step-interpolator representation. Direct-D/A output-signal recovery generally requires a substantial update sample rate f_s to achieve useful amplitude resolution. Since this is an inefficient use of system resources, linear-interpolator and active-filter reconstruction methods offer improved

Figure 9-4 Signal-Reconstruction Model

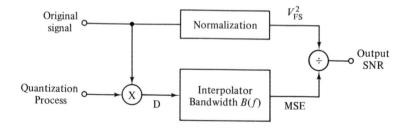

amplitude resolution while conserving sample rate. As a quantitative criterion for evaluating the efficiency of various reconstruction methods we will use the mean-squared error (MSE) derivation of equation (9-3) represented in Figure 9-4. This error is a function of the quantization process originating in the A/D converter (the difference between the original signal and its digital approximation), which is reconstituted by the DAC and dependent upon f_s, signal amplitiude V_s and BW. The merit of various reconstruction methods may then be evaluated with regard to their ability to minimize MSE, thereby providing an improved estimate of the original signal.

The equivalent power density D watts/Hz of the quantization process is obtained with the aid of equation (8-9), and includes normalization by the D/A converter rectangular-equivalent bandwidth of $f_s/2$ Hz obtained from the definite integral of equation (9-4). This insures a valid basis of comparison for the various reconstruction methods given in Table 9-1, for which each amplitude response $A(f)$ has been converted to a rectangular-equivalent interpolator bandwidth $B(f)$. It follows then that equation (9-5) provides the output SNR for the reconstruction method of interest, from which the resolution may be obtained directly with reference to Table 9-2. This equation also includes normalization relative to the signal V_{FS} to permit accommodation of composite signal components less than full scale, such as encountered with harmonic signals. For example, if V_s varies in relation to V_{FS} as a function of frequency, this will be reflected in the SNR and relative output resolution. Equation (9-6) yields the filter time delay in terms of f_c and the number of poles n.

$$\text{MSE} = D \cdot B(f) \qquad (9\text{-}3)$$

$$= \frac{1}{2} \left(\sqrt{\frac{2}{f_s}} \ \frac{V_s \pi BW}{f_s} \right)^2 \cdot B(f) \quad \text{watts}$$

$$B(f) = \frac{1}{2\pi} \int_0^\infty A^2(f) \cdot df \qquad (9\text{-}4)$$

$$= \frac{1}{2\pi} \int_0^\infty \left(\frac{\sin \pi f/f_s}{\pi f/f_s} \right)^2 \cdot df$$

$$= \frac{f_s}{2} \quad \text{Hz (direct D/A)}$$

$$\text{output SNR} = \frac{V_{FS}^2}{\text{MSE}} \qquad (9\text{-}5)$$

$$= \frac{f_s^3 \, V_{FS}^2}{B(f) \, \pi^2 BW^2 \, V_s^2}$$

TABLE 9-1
COMPARISON OF SIGNAL-RECONSTRUCTION METHODS

Signal Type	Reconstruction Method	Amplitude Response, $A(f)$	Interpolator Bandwidth, $B(f)$ (Hz)	Quantization Density, D (watts/Hz)	Filtered MSE (watts)	Output SNR	Time Delay (sec)
DC, sinusoidal, complex harmonic	Direct-D/A	$\dfrac{\sin \pi f/f_s}{\pi f/f_s}$	$\dfrac{f_s}{2}$	$\dfrac{1}{2}\left(\sqrt{\dfrac{2}{f_s}}\dfrac{V_s\pi BW}{f_s}\right)^2$	$\dfrac{1}{2}\left(\dfrac{V_s\pi BW}{f_s}\right)^2$	$\dfrac{2f_s^2 V_{FS}^2}{\pi^2 BW^2 V_s^2}$	0
DC, sinusoidal, complex harmonic	Linear Interpolator	$\dfrac{\sin^2(\pi f/f_s)}{(\pi f/f_s)^2}$	$\dfrac{f_s}{3}$	$\dfrac{1}{2}\left(\sqrt{\dfrac{2}{f_s}}\dfrac{V_s\pi BW}{f_s}\right)^2$	$\left(\sqrt{\dfrac{1}{3}}\dfrac{V_s\pi BW}{f_s}\right)^2$	$\dfrac{3f_s^2 V_{FS}^2}{\pi^2 BW^2 V_s^2}$	$\dfrac{1}{f_s}$
DC, sinusoidal	3-Pole Butterworth	$\dfrac{1}{\sqrt{1+(f/f_c)^6}}$	$\dfrac{f_c}{6}$	$\dfrac{1}{2}\left(\sqrt{\dfrac{2}{f_s}}\dfrac{V_s\pi BW}{f_s}\right)^2$	$\left(\sqrt{\dfrac{f_c}{6f_s}}\dfrac{V_s\pi BW}{f_s}\right)^2$	$\dfrac{6f_s^3 V_{FS}^2}{f_c\pi^2 BW^2 V_s^2}$	$\dfrac{3}{4f_c}$
Complex harmonic	3-Pole Bessel	$\dfrac{1}{\sqrt{1+7.5\left(\dfrac{f}{f_c}\right)^2+3.25\left(\dfrac{f}{f_c}\right)^4+0.16\left(\dfrac{f}{f_c}\right)^6}}$	$\dfrac{f_c}{1.7}$	$\dfrac{1}{2}\left(\sqrt{\dfrac{2}{f_s}}\dfrac{V_s\pi BW}{f_s}\right)^2$	$\left(\sqrt{\dfrac{1.17f_c}{2f_s}}\dfrac{V_s\pi BW}{f_s}\right)^2$	$\dfrac{1.7f_s^3 V_{FS}^2}{f_c\pi^2 BW^2 V_s^2}$	$\dfrac{3}{8f_c}$
Complex harmonic	4-Pole Besselworth	$\dfrac{1}{\sqrt{1+(f/f_c)^8}}$	$\dfrac{f_c}{6.2}$	$\dfrac{1}{2}\left(\sqrt{\dfrac{2}{f_s}}\dfrac{V_s\pi BW}{f_s}\right)^2$	$\left(\sqrt{\dfrac{f_c}{6.2f_s}}\dfrac{V_s\pi BW}{f_s}\right)^2$	$\dfrac{6.2f_s^3 V_{FS}^2}{f_c\pi^2 BW^2 V_s^2}$	$\dfrac{1}{f_c}$

$$\text{filter time delay} = \frac{n}{kf_c} \quad \text{seconds} \qquad (9\text{-}6)$$

where

n = number of poles

k = 4 Butterworth, Besselworth

k = 8 Bessel

Requirements for the design of analog output systems are largely the same as for analog input systems when accuracy is an essential consideration. The choice of update sample rate f_s determines the three system errors of signal aliasing (Table 8-1), average NRZ sin x/x error [equation (8-6)], and reconstruction MSE [equation (9-3)], in increasing order of significance. This latter term provides the recovered-signal output-amplitude resolution by means of equation (9-5) and Table 9-2, although an update rate in excess of the f_s utilized for data conversion obviously cannot further improve resolution. Similarly, an update f_s less than that used for conversion will degrade output resolution indicating that an identical sample and update rate is optimum. In

TABLE 9-2
AMPLITUDE ERROR VERSUS SNR

BINARY EQUIVALENT (BITS)	*AMPLITUDE ERROR (1 LSB) (%FS)*	*OUTPUT SNR (NUMERICAL)*
1	50.0	8.0×10^0
2	25.0	3.2×10^1
3	12.5	1.2×10^2
4	6.2	5.0×10^2
5	3.1	2.0×10^3
6	1.6	8.0×10^3
7	0.8	3.2×10^4
8	0.4	1.2×10^5
9	0.2	5.0×10^5
10	0.1	2.0×10^6
11	0.05	8.0×10^6
12	0.024	3.2×10^7
13	0.012	1.2×10^8
14	0.006	5.0×10^8
15	0.003	2.0×10^9
16	0.0016	8.0×10^9
17	0.0008	3.2×10^{10}
18	0.0004	1.2×10^{11}
19	0.0002	5.0×10^{11}
20	0.0001	2.0×10^{12}

addition, output signals that are a composite of more than one sampled-data signal require an update f_s equal to that of the highest sample rate included, to insure signal aliasing does not occur. The linear interpolator requires a sample-hold, an integrator with a gain function $1/RC$ equal to $f\omega$, plus an output summer. It does provide increased sideband attenuation in comparison with direct-D/A signal recovery, but it also exhibits a greater gain error in the region where the recovered signals normally reside, as shown in Figure 9-3.

With reference to Table 9-1 and its tabulation of interpolator bandwidth functions $B(f)$, linear interpolation is comparable in effectiveness but not as simple in its mechanization as active filtering for signal reconstruction. A further consideration is that signal-power spectra typically have long time-average properties, such that filters are especially efficient in minimizing MSE as filter bandwidth decreases. The three-pole Butterworth lowpass filter is a notable example for the reconstruction of dc and sinusoidal type signals. The need for the Bessel and Besselworth characteristics is to minimize filter component error in the reconstruction of complex harmonic signals. The use of active filters for signal recovery draws upon developments of Chapter 4 for the necessary circuit design and component error considerations. We shall examine the interrelationship of these quantities in detail in the filter signal-reconstruction design examples that follow.

Possible errors associated with digital processor operations on these signals, such as register overflow and roundoff noise, are not represented in these results. However, their magnitude is usually small and in the range of $\pm\frac{1}{2}$ LSB effects previously considered with A/D conversion. A possible exception to the use of filter reconstruction methods may be encountered in automatic control systems, where the filter time delay is undesirable within feedback loops. Fortunately, these signals frequently are narrowband, requiring only a modest update rate, and usually interface with actuators capable of only a limited frequency response which assist the signal-reconstruction function. If the terminating frequency response can be characterized, then it too may be developed into a quantitative reconstruction method by conversion to an interpolator bandwidth $B(f)$ using equation (9-4) and applying Tables 9-1 and 9-2.

9-2 SIGNAL-RECONSTRUCTION APPLICATIONS AND ERROR ANALYSIS

MSE criteria for signal reconstruction are applied to both the low- and high-data-rate conversion systems of Section 8-3. Active lowpass filters are used in the interest of efficient and economical signal recovery. Twelve-bit voltage-output D/A converters are employed for both reconstruction systems in order to minimize this component error. The first-order rolloff for the complex harmonic signal between 100 Hz and 1 kHz is accommodated in the MSE and output SNR equations by appropriate amplitude scaling. A four-pole Bessel-worth filter is utilized for the high-data-rate recovery system and is specified for the signal bandwidth of 1 kHz in accordance with Table 4-9. A three-pole Bessel mechanization is inadequate because of its marginal efficiency in this application plus its moderate component error contribution. A three-pole Butterworth filter, however, is adequate for recovery of the 0.1-Hz BW thermocouple signal in the low-data-rate system. Signal aliasing can occur in a data-recovery system if the selection of update rate f_s results in undersampling. However, noise aliasing cannot occur such as in a data-conversion system because the total reconstructed spectrum is specified *a priori* in terms of the input digital signal.

Solution of equation (9-5) with the same sample rates used for the conversion of these signals, 10 kHz and 10 Hz respectively for the high- and low-data-rate systems, yields SNR values resulting in an output amplitude resolution equivalent to seven binary bits (0.8%) for the complex signal and eight binary bits (0.4%) for the thermocouple signal. Realized is a two-bit resolution improvement over the five-bit step-interpolator representation of the data in its sampled form for the high-data-rate system, and a three-bit improvement for the low-data-rate system, where an earlier filter cutoff is permissible than in the high-data-rate system. The application of equation (9-5) for the output resolution of interest from Table 9-2, coordinated with an appropriate recovery method from Table 9-1, may therefore be considered in the method-of-approach for minimizing the conversion-system sample rate in all analog-I/O-system designs. These results may be validated by substitution of the applicable recovery-system values into equation (9-5) for direct-D/A signal reconstruction. The resulting amplitude resolution error from Table 9-2 will be identical to the conversion-system step-interpolator amplitude resolution determined by equation (8-10) for both systems.

The details of these signal-reconstruction systems are presented in the calculations that follow and are summarized in Table 9-3. Note that the source errors represented by the high- and low-data-rate conversion systems of Chapter 8 are combined in an rms fashion with the reconstruction-system errors. This includes substitution of the smoothed output amplitude resolution for that of the step-interpolator sampled-data resolution, which is responsible for improvement in the data-conversion system accuracy from approximately five bits to a reconstruction-system output accuracy equivalent to six bits at the maximum 1-kHz bandwidth for the high-data-rate system. For the low-data-rate system, an output accuracy improvement from five bits to seven bits is achieved at the maximum 0.1-Hz bandwidth. These results clearly demonstrate the influence of the signal recovery system on the selection of a conversion and reconstruction f_s for achieving the input-to-output accuracy of interest.

TABLE 9-3

ANALOG OUTPUT-SYSTEM ERRORS AT FULL BANDWIDTH

SYSTEM	LOW DATA RATE		HIGH DATA RATE	
Signal Type	DC and Sinusoidal		Complex Harmonic	
Component	Error (%FS)	Comment	Error (%FS)	Comment
D/A	0.04	12-bit voltage output	0.04	12-bit voltage output
Filter	0.2	3-pole Butterworth 0.16 Hz f_c	0.2	4-pole Besselworth 1.6 kHz f_c
Resolution	0.4	Output SNR $= 3.8 \times 10^5$	0.8	Output SNR $= 4.0 \times 10^4$
Sin x/x	0.009	Average attenuation	0.82	Average attenuation
Aliasing	0.0	For $\dfrac{f_s}{BW} = 100$	0.0	For $\dfrac{f_s}{BW} = 10$
RMS reconstruction	0.45	8-bit accuracy	1.16	7-bit accuracy
RMS source	0.87	Conversion system error at 0.1-Hz BW maximum minus step-interpolator resolution $(3.22\%^2 - 3.1\%^2)$	1.49	Conversion system error at 1-kHz BW maximum minus step-interpolator resolution $(3.44\%^2 - 3.1\%^2)$
RMS total	0.98%	7-bit accuracy	1.89%	6-bit accuracy

At one-tenth bandwidth the low-data-rate system total error is reduced to 0.75%, or seven-bit equivalent output accuracy, owing primarily to improvement in output amplitude resolution. However, further bandwidth reduction provides negligible additional improvement, and the 0.74% residual output error is essentially identical to the 0.73% value for the data-conversion system. Repeating the calculations for the high-data-rate recovery system results in the same trend exhibited by its conversion system, whereby at one-tenth bandwidth the output error is only slightly changed from its full bandwidth value to 1.49% because of the constant amplitude resolution of the harmonics down to one-tenth bandwidth. Below one-tenth bandwidth the high-data-rate output error decreases asymptotically to its residual value of 1.30%. Both of these error relationships, as illustrated in Figure 9-5, indicate less overall variance as a function of bandwidth than the conversion-system errors of Figure 8-9. This is essentially attributable to the efficiency of active filtering in minimizing MSE. For eight-bit digital processors eight data lines can be connected to the 12-bit D/A converters with essentially no change in these results. This is because the four omitted LSB's of the D/A represent less than the residual error of either system shown by Figure 9-5.

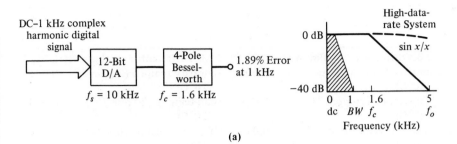

(a)

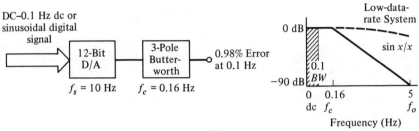

(b)

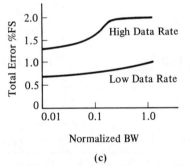

(c)

Figure 9-5 Analog Output Systems: (a) High Data Rate, (b) Low Data Rate, (c) Total Error

LOW-DATA-RATE SYSTEM	HIGH-DATA-RATE SYSTEM
Sources	
DC-0.1 Hz thermocouple signal converted with 3.22% error at a 10-Hz sample rate providing five-bit (3.1%) step-interpolator resolution. V_{FS} to 0.1-Hz BW.	DC-1 kHz complex harmonic signal converted with 3.44% error at a 10-kHz sample rate providing five-bit (3.1%) step-interpolator resolution. V_{FS} to 100 Hz with rolloff to 0.1 V_{FS} at 1 kHz BW.

LOW-DATA-RATE SYSTEM		*HIGH-DATA-RATE SYSTEM*	
D/A Converters			
12-Bit differential nonlinearity	0.012%	12-Bit differential nonlinearity	0.012%
Linearity tempco (2 ppm/°C) (20°C)	0.004	Linearity tempco (2 ppm/°C) (20°C)	0.004
Gain tempco (20 ppm/°C) (20°C)	0.040	Gain tempco (20 ppm/°C) (20°C)	0.040
Offset tempco (5 ppm/°C) (20°C)	0.010	Offset tempco (5 ppm/°C) (20°C)	0.010
RMS total$_{\%FS}$	0.04%	RMS total$_{\%FS}$	0.04%
Filters			
3-Pole Butterworth lowpass active filter, amplitude error 0.2% for dc and sinusoidal signals. $f_c = 0.16$ Hz (Table 4-9).		4-Pole Besselworth lowpass active filter, amplitude error 0.2% for complex harmonic signals. $f_c = 1.6$ kHz (Table 4-9).	

Resolution

Output SNR

$$= \frac{6 f_s^3 V_{FS}^2}{f_c (\pi BW)^2 V_s^2} \qquad (9\text{-}5)$$

$$= \frac{(6)\,(10\text{ Hz})^3\,(1.0)^2}{(0.16\text{ Hz})(\pi \cdot 0.1\text{ Hz})^2 (1.0)^2}$$

$$= 3.8 \times 10^5$$

$\epsilon_{\text{resolution}\,\%FS}$
$\quad = 0.4\%$ (Table 9-2)

Output SNR

$$= \frac{6.2 f_s^3 V_{FS}^2}{f_c (\pi BW)^2 V_s^2} \qquad (9\text{-}5)$$

$$= \frac{(6.2)(10\text{ kHz})^3\,(1.0)^2}{(1.6\text{ kHz})(\pi \cdot 1\text{ kHz})^2 (0.1)^2}$$

$$= 4.0 \times 10^4$$

$\epsilon_{\text{resolution}\,\%FS}$
$\quad = 0.8\%$ (Table 9-2)

Sin x/x

Average NRZ sin x/x error

$$= \frac{1}{2}\left(1 - \frac{\sin \pi BW/f_s}{\pi BW/f_s}\right) \cdot 100\% \quad (8\text{-}6)$$

$$= \frac{1}{2}\left(1 - \frac{\sin \pi/100}{\pi/100}\right) \cdot 100\%$$

$$= 0.009\%$$

Average NRZ sin x/x error

$$= \frac{1}{2}\left(1 - \frac{\sin \pi BW/f_s}{\pi BW/f_s}\right) \cdot 100\% \quad (8\text{-}6)$$

$$= \frac{1}{2}\left(1 - \frac{\sin \pi/10}{\pi/10}\right) \cdot 100\%$$

$$= 0.82\%$$

Aliasing

Signal aliasing
essentially zero
for f_s/BW = 100
(Table 8-1)

Signal aliasing
essentially zero
for f_s/BW = 10
(Table 8-1)

LOW-DATA-RATE SYSTEM	HIGH-DATA-RATE SYSTEM
Time Delay	
Recovery filter time delay $3/(4f_c) = 4.7$ s for 0.16-Hz cutoff three-pole Butterworth lowpass filter.	Recovery filter time delay $1/f_c = 0.625$ ms for 1.6-kHz cutoff four-pole Besselworth lowpass filter.

9-3 SIGNAL DISTRIBUTION AND TRANSMISSION

The preceding sections developed the design of recovery systems for the reconstruction of analog signals from digital data. The methods presented implied the use of a D/A converter per channel to allow design optimization for the specific characteristics of each signal. This preferred data-distribution technique is illustrated in Figure 9-6 together with an alternative arrangement, which employs a single DAC followed by a sample-hold per channel. The sample-hold per channel, an earlier signal-distribution approach used when D/A converters were relatively expensive, imposed the requirement of frequent updates to minimize output error even with slowly varying or static data. Further, the lowpass response of the output actuator or terminating device usually was relied upon to provide signal-reconstruction smoothing. Today, however, inexpensive D/A converters with input registers and output filters offer improved performance.

Some analog output applications do not involve intrinsic signal reconstruction but are classifiable as programmable functions. Digitally programmable power supplies are included in this category. Figure 9-7 shows an output channel with a programmable multiplying DAC and four-terminal power amplifier. This analog output channel could be incorporated in a computer-driven automatic test system with the digital input representing a computed test limit and V_i an externally supplied analog test signal. The four-terminal amplifier connection compensates for output line-voltage drop by accurately sensing V_o across R_L. The distribution of analog voltages poses fewer problems than acquisition owing to the higher signal levels involved. However, direct-wire transmission distance is limited to about 100 feet for both, the limiting factors being signal attenuation with distribution and interference pickup with acquisition.

Distributing a current replica of a signal is advantageous when distances exceed 100 feet, because attenuation errors are eliminated and the low-impedance circuit minimizes induced interference. Figure 9-8 describes practical voltage- and current-converter circuits which can accommodate line

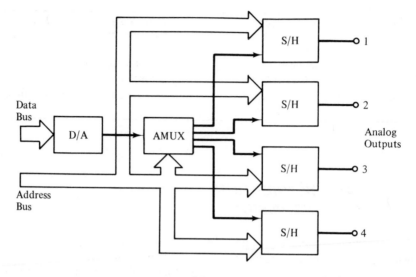

(a)

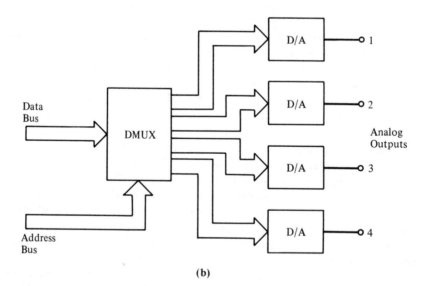

(b)

Figure 9-6 Data Distribution Methods

resistances between 0 and 249 ohms without accuracy degradation. The current output amplifier must be capable of delivering 20 mA, which can be met by devices such as the Burr Brown 3268/14. Analog Devices offers complete modular 4- to 20-mA voltage-to-current converters in their 2B20

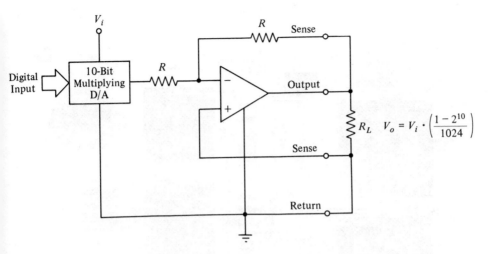

$$V_o = V_i \cdot \left(\frac{1 - 2^{10}}{1024} \right)$$

Figure 9-7 Multiplying DAC with Four-Terminal Amplifier Output

Figure 9-8 Voltage and Current Converters

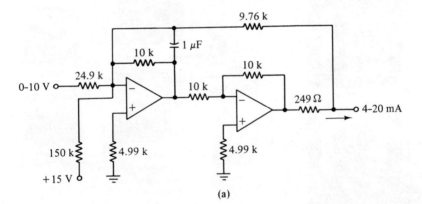

(a)

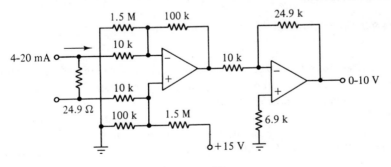

(b)

239

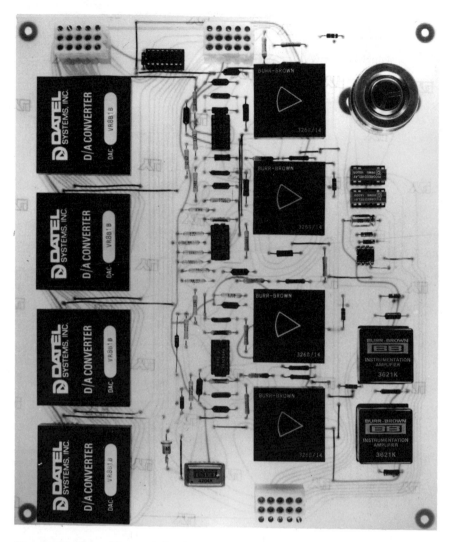

Figure 9-9 Four-Channel Analog Output Board

Series devices. Isolated voltage or current outputs may be achieved by insertion of an isolation amplifier, generally preceding the final stage. Practical current-loop transmission distances extend to 1000 feet, and both current and voltage signals benefit from shielded cable. Grounding and shielding practice for analog output systems is essentially the same as that described for analog input systems in Section 6-1. In addition, protection from fault conditions can be provided by a series-connected slo-blo fractional-ampere fuse at the output terminals with a GE ±10-V MOV device shunted to ground on the inside of this connection. Figure 9-9 shows a four-channel analog output system with current converters for interfacing a digital computer to process control apparatus.

When data-acquisition or signal-distribution distances exceed 100 feet from the terminating point, the signal-transmission methods of Table 9-4 may be used for ranges to 10 miles. These methods service most requirements without complicated mechanizations or substantial cost, and some employ digital signalling techniques because of their minimal signal-to-noise requirement for accurate transmission in comparison with analog methods. Distances for a majority of remote-signal-transmission applications are less than a mile. MODEM eliminators represented by the Intersil REMDACS and Analog Devices μ MAC-4000 are representative of the flexible, self-contained signal transmission subsystems available. Depending upon the options selected, these modules may also include signal-conditioning and conversion functions in addition to parallel/serial commutation, and they are compatible with the conventional RS232C serial-digital electrical standard presented in Table 9-5.

The mechanization of a basic MODEM eliminator transmitter and receiver is shown by Figure 9-10. Interconnection requires a properly terminated balanced-wire pair to cancel signal reflections and for operation of the differential line drivers and receivers. These devices usually will accommodate common-mode line noise up to ±3 V. However, noise is not the principal limitation on transmission distance or accuracy. Both the line bandwidth and

TABLE 9-4
SIGNAL-TRANSMISSION METHODS

METHOD	SIGNAL	DISTANCE	SPEED	APPLICATION
Reconstruction circuits	Analog voltage	100 ft	1 MHz	Signal-distribution purposes
Current converters	Analog current	1000 ft	10 kHz	Process-control standard
Line drivers	Serial digital	1 mile	600 bps	MODEM eliminator
Limited-distance MODEM	Binary-modulated analog signals	10 miles	1200 bps	Dedicated-line transmission
Radio MODEM	Binary-modulated analog signals	10 miles	300 bps	Wireless 11-meter-band transmission

TABLE 9-5
LINE ELECTRICAL STANDARDS

| PARAMETER | RS232C | | RS422 | |
	Driver	Receiver	Driver	Receiver
ONE	− 5 V min −15 V max	−3 V max	A negative	A < B
ZERO	+5 V min +15 V max	+3 V max	B positive	A > B
Impedances	Unspecified	3 k min 7 k max	< 100Ω balanced	> 4 k
Max levels	±500 mA	±25 V	150 mA	±6 V

bit-transmission rate affect the bit-timing uncertainty, or intersymbol interference, at the receiving end which determines the error rate achieved. The line bandwidth decreases with increasing distance as a result of cumulative shunt capacitance and series line inductance. The Nyquist rate R defines the maximum allowable bit rate for minimum intersymbol interference in terms of this bandwidth by equation (9-7), where the received-bit risetime t_r is a

Figure 9-10 MODEM Eliminator Transmitter-Receiver

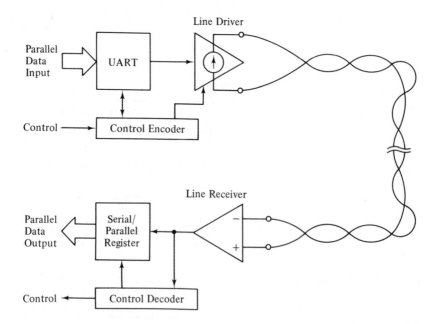

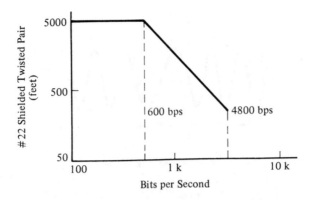

Figure 9-11 Maximum Line Length Versus Bit Rate

readily obtained measurement proportional to line bandwidth. For example, a receiving-end bit risetime measurement of 500 μs indicates a maximum signaling rate of 2 kbps for an error-rate probability per bit of 10^{-4} or better. Figure 9-11 shows allowable line lengths versus bit rates for this error probability.

$$R \leq 2BW \quad \text{bps} \qquad (9\text{-}7)$$

$$\leq \frac{1}{t_r}$$

Data transmission over distances greater than about a mile requires additional system complexity to overcome the previous line-bandwidth distance-limiting factor. A MODEM provides a solution by encoding the digital information and then modulating it by a method suitable for efficient transmission over the available bandwidth of an analog wire or radio channel. Limited-distance MODEMs bring together digital signals and analog channels, which are usually narrowband, and typically employ frequency-shift-keyed (FSK) modulation with asynchronous operation in either half-duplex (reversible-direction) or full-duplex (simultaneous two-way) modes. Advantages of FSK modulation, shown in Figure 9-12, include an error rate that is essentially independent of signal amplitude, equal per-digit error probabilities for a Mark and Space, and simple noncoherent detection without need to process the carrier. Limited-distance MODEMs provide greater range than does the previous direct-binary transmission method because FSK tones have narrower bandwidth requirements than do baseband binary pulses used by line drivers and receivers. However, intersymbol interference, which results

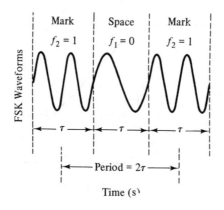

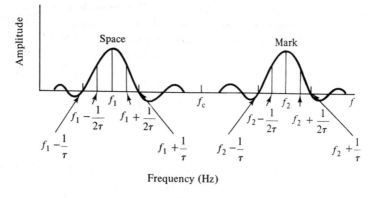

Figure 9-12 FSK Time- and Frequency-Domain Signal Set

from spreading of the signal envelopes of Figure 9-12 owing to the contracting bandwidth of extended wire lines or to cumulative signal attenuation and additive noise in the case of a radio MODEM, eventually limits the range of low-error transmission with limited-distance MODEMs also.

MODEMs consist of two functional systems described in Figure 9-13. The modulator accepts binary signals and converts them into analog audio-frequency signals for transmission, and an output bandpass filter eliminates out-of-band harmonic energy. Input line-conditioning filters may be provided to compensate for the attenuation and delay distortion encountered over wire links described by Figure 9-14. The demodulator reconverts the audio-

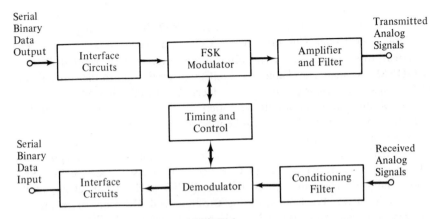

Figure 9-13 Limited Distance MODEM

frequency signals back to binary signals. In addition to intersymbol interference, transmission errors on long wire lines are also caused by random noise bursts lasting to 50 ms. For the typical slow-speed limited-distance MODEM, the bit transmission rate usually does not exceed 1.2 kbps to avoid intersymbol interference. Therefore, periodic burst noise is not a significant consideration. However, higher-speed MODEMs routinely include an error-control system that is also effective against burst errors. This usually is achieved with a redundant encoding scheme that permits accurate data reconstruction at the receiver with an error burst up to 50 bits in length.

Commercial limited-distance FSK MODEMs typically have a 1-kHz

Figure 9-14 Wire Line Delay and Attenuation

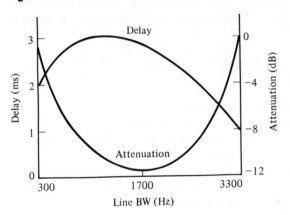

separation between Mark and Space frequencies, respectively 2200 and 1200 Hz with a 1700-Hz carrier frequency, and an RS232C interfacing standard as available from Codex Corporation and General Electric Data Products. The practical line speeds for serial-binary line drivers and limited-distance MODEMs, as shown in Table 9-4, limit their application to the transmission of low-data-rate signals of only tens of Hertz bandwidth, depending upon the digital wordlength or amplitude resolution required. Note that the MODEM eliminator of Figure 9-10 may also be used to interface parallel binary data to a limited-distance MODEM.

Carrier-frequency allocations for radio remote-control purposes are covered under Part 95 of the FCC Rules. This Class C service for fixed-station nonvoice communications includes 27.255 MHz (Channel 23) of the citizen's band service. Consequently, a commercial CB transceiver may be utilized as a radio data MODEM with the addition of an FSK modulator and demodulator as described by Figure 9-15. The only qualification is antenna attachment directly to the transmitter with a gain not exceeding that of a half-wave dipole. A noncoherent FSK signal is generated by the Exar 2207 voltage-controlled oscillator and detected by an Exar 567 phase-locked loop tone decoder in half-duplex operation with this example.

The performance of a radio MODEM is constrained both by intersymbol interference and the SNR associated with the receiver. The primary limitation is SNR, however, which can be characterized by equation (9-8), where the receiver-processing-gain improvement of input SNR_i is determined by the ratio of the front end bandwidth B_{f_e} to that of the detector B_d. Since the first sidebands appear at $\pm 1/2\tau$ Hz on either side of the Mark and Space frequencies as shown in Figure 9-12, a 300-bps data rate requires a minimum detector bandwidth of 300 Hz (± 150 Hz). With a typical receiver B_{f_e} of 10 kHz the available processing gain is 33.3, and the required SNR_i is 0.51 (-2.9 dB) for a 10^{-4} error rate. Reliable transmission has been obtained with this FSK radio MODEM over a 10-mile distance in an urban environment.

$$\text{error rate} = \frac{1}{2} \exp \left(-\frac{SNR_i}{2} \cdot \frac{B_{f_e}}{B_d} \right) \qquad (9\text{-}8)$$

Industrial process control links and communications between distributed microcomputers increasingly are adopting the flexibility and features offered by the bus structure defined in MIL–STD–1553B. This time-division multiplexed system provides for bidirectional serial information transmission in the half-duplex mode among addressed transceivers interconnected by a single twisted shielded-pair of conductors. The bus is a transformer-coupled, fault-isolated asynchronous transmission line as illustrated by Figure 9-16. ILC Data Device Corporation manufactures components for this phase-

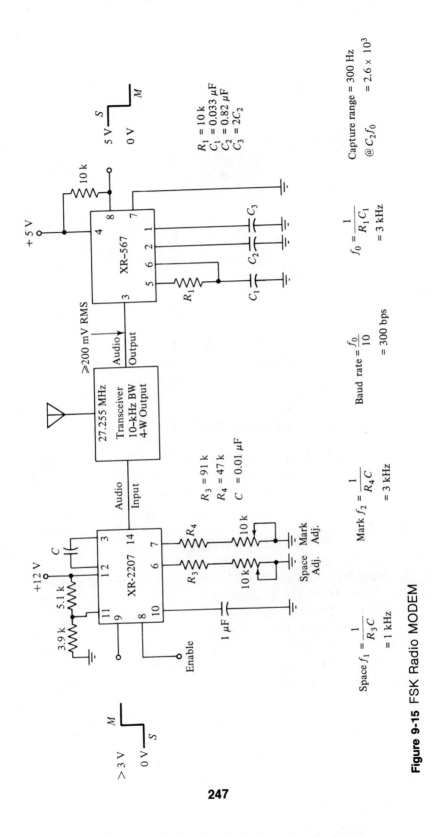

Figure 9-15 FSK Radio MODEM

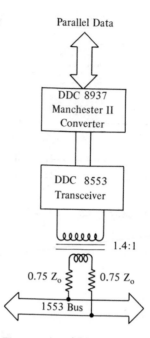

Figure 9-16 1553 Serial
Biphase MODEM Bus

modulated MODEM including a Manchester II converter which encodes parallel binary data into biphase serial data. A 1-MHz bit-rate information flow is possible with this system.

REFERENCES

1. H. NYQUIST, "Certain Topics in Telegraph Transmission Theory," *Trans. AIEE*, Vol. 47, April 1928.

2. G. LAPIDUS, "Transmitting Data Pulses Over Short Distances," *Data Communications*, July/August 1974.

3. "Low-Speed MODEM Fundamentals," Application Note AN-731, Motorola Incorporated, Phoenix, Ariz., 1974.

4. *The Communications Handbook*, Microdata Corporation, Irvine, Calif., 1973.

5. A. B. CARLSON, *Communication Systems*, 2nd Ed., McGraw-Hill, New York, 1975.

6. H. STARK and F. TUTEUR, *Modern Electrical Communications*, Prentice-Hall, Englewood Cliffs, N.J., 1979.

7. M. SCHWARTZ, *Information Transmission Modulation and Noise*, 3rd Ed., McGraw-Hill, New York, 1980.

8. R. ZIEMER and W. TRANTER, *Principles of Communications*, Houghton Mifflin, Boston, 1976.

9. "Transmission Line Characteristics," National Semiconductor Application Note AN-108, Santa Clara, Calif.

10. D. H. AXNER, "Compatibility and Transmission Modes Steer MODEM Selection," *Data Comm User*, May 1976.

PROBLEMS

9-1. Verify five-bit output step-interpolator resolution for the high-data-rate system at full bandwidth for direct-D/A signal recovery with the aid of Tables 9-1 and 9-2. Then include a three-pole Bessel reconstruction filter and determine the output-resolution improvement.

9-2. Evaluate the error terms of Table 9-3 for the low data-rate system at one-tenth the 0.1-Hz signal bandwidth to verify that point on the error curve of Figure 9-5.

9-3. Design a premium signal-recovery channel for the dc−1-kHz undefined-type signal of Problem 8-6 that minimizes overall amplitude error. Show the components and the output-frequency spectrum, and calculate the input-to-output rms error budget including the conversion-system source error determined at the same sample rate used in the solution of Problem 8-6.

9-4. Binary data are sent at 10 kbps over a twisted-pair line having a bandwidth of 6.7 kHz. Determine whether intersymbol interference is a problem.

9-5. A 1200-bps data rate generates an FSK signal set with Mark and Space frequencies of 3.6 and 1.2 kHz. Determine the transmission bandwidth required for this signal set and sketch its frequency spectrum.

9-6. An operational-amplifier duplexer is used to couple an FSK MODEM to a telephone line as shown. Select resistor values that maximize gains A_1 and A_2 while minimizing A_3. Determine the values of these gains.

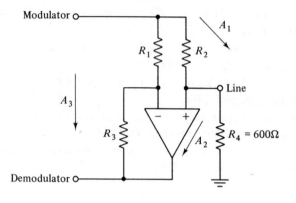

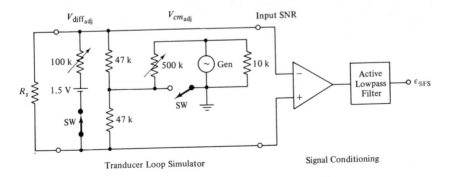

experiments in signal acquisition

Set up the transducer simulator as shown, where the V_{cm} may be dc, coherent ac, or a random noise generator. The source resistance R_s may be specified between a few tens of ohms and a few kilohms. Alternately adjust the floating V_{diff} to 10 mV dc and the V_{cm} to 1 V rms 60 Hz or random interference at the transducer loop output. Then calculate the signal-conditioning input SNR using equation (6-3) at these voltage values, where the rms reading is multiplied by 1.127 with random noise for an average responding voltmeter calibrated for sine waves.

Select an instrumentation amplifier and calculate its component error by the methods of Table 3-5 for an $A_{v_{\text{diff}}}$ of 100. Then choose an appropriate

Tranducer Loop Simulator   Signal Conditioning

three-pole unity-gain active filter from Table 4-9 and perform the appropriate signal-quality calculations for a 1-Hz f_c using equations (6-4) through (6-8), Figure 6-6, and Table 6-2. Determine the signal-conditioning channel error $\epsilon_{\%FS}$ for the 1-V output signal, where the transducer is considered error-free for this exercise, and also the equivalent binary-bit accuracy.

Connect the instrumentation amplifier to the transducer simulator and verify the amplifier output SNR signal quality by alternately switching off the V_{diff} and V_{cm} signal sources. Then implement the chosen active filter type with the aid of Chapter 4 and verify the filter output SNR in the same manner. Repeat for both coherent and random interference if a random-noise generator is available.

appendix b

experiments in data conversion

An A/D conversion system is illustrated, using the voltage-to-frequency and gated-counter technique described in Chapter 7. Construct this converter using a 0−10 V/0−10 kHz V/F converter, and determine the required gating interval τ to achieve four-bit conversion for a 10-V input signal. Also, compare

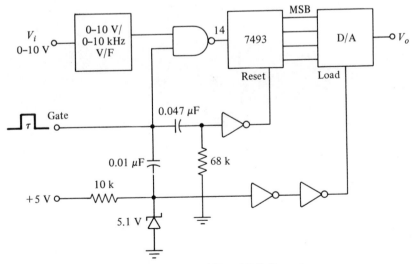

Four-bit Voltage/Frequency A/D and D/A Converters

the LSB amplitude obtained with the value shown in Table 8-3. The error of this conversion system is dominated by its step-interpolator amplitude resolution error.

Connect the four MSBs of a D/A converter to the output of this A/D converter, and apply a 0- to 10-V triangular signal V_i at a period of ten times the gating interval τ. Observe the DAC output and verify proper signal quantization. Apply V_o to Channel One and V_i to Channel Two of an oscilloscope with an "invert and add" feature to obtain the quantization error described by Figure 7-16. Plot the peak-to-peak quantization error as the signal period is reduced to τ.

experiments in signal recovery

Consult Table 4-9 to determine the appropriate three-pole recovery filter type and its f_c to reconstruct a 0- to 10-V sinusoidal signal V_i centered at 5 V for the unipolar A/D converter. Apply the methods of Chapter 9, including equation (9-5) and Tables 9-1 and 9-2, to calculate the expected output resolution for a signal period $(1/BW)$ 100 times the required gating interval τ determined in the data-conversion experiment.

Construct this filter with the aid of Chapter 4 and compare V_o with the original V_i. Then move the sinusoidal signal source to the transducer loop simulator for an acquisition-conversion-recovery experiment, with the necessary scaling adjustments to provide a 0- to 10-V signal at the signal-conditioning output. A 3-2 prong line-cord adapter may improve the isolation of this differential signal source. Adjustments in the signal-conditioning filter may be required to properly accommodate the spectral occupancy requirements of the signal source.

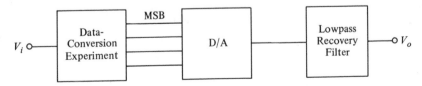

Output-Signal Reconstruction

review of decibels

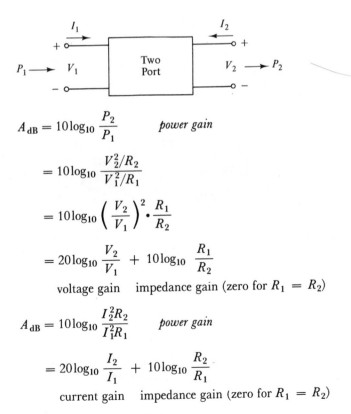

$$A_{dB} = 10\log_{10}\frac{P_2}{P_1} \qquad \textit{power gain}$$

$$= 10\log_{10}\frac{V_2^2/R_2}{V_1^2/R_1}$$

$$= 10\log_{10}\left(\frac{V_2}{V_1}\right)^2 \cdot \frac{R_1}{R_2}$$

$$= 20\log_{10}\frac{V_2}{V_1} + 10\log_{10}\frac{R_1}{R_2}$$

voltage gain   impedance gain (zero for $R_1 = R_2$)

$$A_{dB} = 10\log_{10}\frac{I_2^2 R_2}{I_1^2 R_1} \qquad \textit{power gain}$$

$$= 20\log_{10}\frac{I_2}{I_1} + 10\log_{10}\frac{R_2}{R_1}$$

current gain   impedance gain (zero for $R_1 = R_2$)

index